Dictionary
for
Clinical Trials

To Nikki, Anya *and* Huw

Dictionary for Clinical Trials

Simon Day

Medical Department, Leo Pharmaceuticals,
Princes Risborough, UK

JOHN WILEY & SONS, LTD

Chichester · New York · Weinheim · Brisbane · Singapore · Toronto

Other Wiley Editorial Offices

John Wiley & Sons, Inc., 605 Third Avenue,
New York, NY 10158-0012, USA

WILEY-VCH Verlag GmbH, Pappelallee 3,
D-69469 Weinheim, Germany

Jacaranda Wiley Ltd, 33 Park Road, Milton,
Queensland 4064, Australia

John Wiley & Sons (Asia) Pte Ltd, 2 Clementi Loop #02-01,
Jin Xing Distripark, Singapore 129809

John Wiley & Sons (Canada) Ltd, 22 Worcester Road,
Rexdale, Ontario M9W 1L1, Canada

Library of Congress Cataloging-in-Publication Data

Day, Simon.
    Dictionary for clinical trials / Simon Day.
        p.   cm.
    Includes bibliographical references.
    ISBN 0-471-98611-9 (cased : alk. paper).—ISBN 0-471-98596-1 (paper : alk. paper)
    1. Clinical trials—Dictionaries.   I. Title.
    [DNLM: 1. Clinical Trials dictionaries.   QV 13 D275d 1999]
R853.C55D39   1999
610'.7'2—dc21
DNLM/DLC
for Library of Congress   99–11151
                                                                                      CIP

British Library Cataloguing in Publication Data

A catalogue record for this book is available from the British Library

ISBN 0-471-98611-9 (cased)
ISBN 0-471-98596-1 (paper)

Typeset in 9/10½pt Times from the author's disks by Vision Typesetting, Manchester
Printed and bound in Great Britain by Antony Rowe Ltd., Chippenham, Wiltshire
This book is printed on acid-free paper responsibly manufactured from sustainable forestry, in which at least two trees are planted for each one used for paper production

Preface

It is now fifty years since the British Medical Research Council published the results of a trial entitled 'Streptomycin treatment of pulmonary tuberculosis' (*British Medical Journal*, 30th October 1948, pages 769–782). That study is widely regarded as the first randomised clinical trial. Earlier examples of nonrandomised studies are cited, notably that of J Lind (*A Treatise on the Scurvy*, 1753). Despite such a history and the enormous numbers of trials conducted and published in the last twenty or so years, many people do not consider 'clinical trials' as a discipline in its own right and, as such, the breadth of terms that should be covered in a dictionary of this kind is not well defined. Ultimately, the choice of entries is a personal one, guided by experiences of what I have had to learn and what my colleagues in various specialities of the clinical trials spectrum have struggled to understand. Additionally I have trawled clinical trial protocols, reports, regulatory guidelines and published manuscripts to try to cover the majority of terms that are likely to be encountered. A lot of the terminology of clinical trials is statistical: terms used for the design (blocks, randomisation, stratification) and for the analysis (confidence interval, *P*-value, survival analysis, *t* test, to list but a few). I make no apology for the high proportion of statistical terms: those are usually the ones that are least well understood. Overall though, the content is broad and it is very difficult to summarise what is covered.

It is almost as difficult to summarise what isn't covered. This is not a dictionary of medical terms, of statistical terms, of epidemiological, ethical or data management terms. It does, however, contain elements of all those disciplines, the first three in particular. Many of the epidemiological terms included would not ordinarily be found in a clinical trial protocol or report; however, in the discussion of whether a clinical trial is appropriate for answering a particular medical question, or in discussion of trial results alongside other sources of evidence, the issue of other approaches such as case-control studies and cohort studies are likely to be discussed. I have not included specific diseases (a medical dictionary would be more appropriate) or names of clinical rating scales but I have included a variety of medical terms that are frequently assumed to be understood (terms such as acute, chronic, subcutaneous, etc.) Abbreviations are not included, except in the few instances where a term is better known by its acronym than by its full name (COSTART and MedDRA are obvious examples). Nor are the names of professional or scientific societies, research institutions or regulatory authorities included.

The intended readership for this dictionary are all those people who work with clinical trial protocols and reports or who otherwise need to understand the use of language in this specialist area. Such a readership includes clinical trialists (those people who actually carry out the various administrative, clerical and scientific aspects), those who sit on ethics committees, those who work in regulatory departments or grant awarding bodies, doctors, nurses, pharmacists (and patients) reading clinical trial reports, and so on. Trials sponsored by the pharmaceutical industry, as well as those conducted by academic institutions or by small groups of enthusiasts, all fall within the scope of this work as do community-based intervention studies, vaccine trials, studies of medical practice and medical devices. Necessarily, many entries will be more relevant to some types of trials and trialists than to others. I hope the coverage is adequate without being too cumbersome.

The style of explanations and definitions is aimed at being pragmatic and readable rather than purist. Pre-existing definitions (often in regulatory guidelines) have not necessarily been faithfully reproduced, although care has been taken to incorporate the essential meaning from relevant guidelines. As an example, the term 'adverse event' has a very specific definition within the International Conference on Harmonisation although the explanation given here is a little more brief. Further examples of pragmatism abound in the explanations of some statistical terms. Many statisticians may challenge the correctness of my explanations of analysis of covariance, Bayesian statistics or P-value, for example: I apologise to them in advance but hope that the explanations I have given will help those readers who understand little or nothing of such terms to at least gain a rough and ready grasp of their meaning. Similarly, 'ethics' is covered in a mere two lines: there are other related entries but the aim is to get the essential meaning across. Full and complete explanations of all the terms included would mean this work taking on the scale of a series of text books and that is not the intention. I hope that the explanations given here, put in the context where the word or expression has arisen, will allow most readers to unravel most uncertainties.

In my defence over accuracy and quality control I can claim that every single entry has been reviewed by a variety of my colleagues; and in their defence I acknowledge that every single error, discrepancy and inconsistency remains my responsibility.

The Ground Rules

The following is a brief guide to what's in and what's not in, and rules for cross-referencing related or alternative terms.

In general, *study* is used rather than *trial* except where the distinction is helpful (strictly speaking, study encompasses trial but many types of study will not be trials). Similarly, *trial* is taken to mean *clinical trial*. For example, *acute study* is listed, but not *acute trial* or *acute clinical trial*.

Phrases may sometimes be abbreviated but, I hope, without causing any difficulty in finding them. For example, *adaptive design* should be taken to encompass *adaptive trial design* and *adaptive clinical trial design*.

Where alternative terms may be used interchangeably I have tried to pick the most common term to define and its synonyms will simply direct you there with the symbol $\approx$. For example, *alpha error* simply says '$\approx$ *type I error*' (where an explanation is given). The most important terms used within the definitions of other terms are emboldened, as are references to contrasting terms ($\Leftrightarrow$ . . .) and related terms ($\Rightarrow$ . . .). I hope that sometimes giving indication of contrasting or related terms may help understanding. It is inevitable, however, that some definitions will be circular: *active control* contrasts with ($\Leftrightarrow$) *placebo control*; *placebo control* contrasts with ($\Leftrightarrow$) *active control*. Ultimately, just as with all dictionaries, all definitions must use the terms herein to explain other terms and the circularity becomes inevitable.

Bibliography

There is a variety of books written about clinical trials and several other dictionaries and glossaries that may prove helpful in defining terms and clarifying their use. The following titles have proved particularly helpful in compiling this dictionary and may serve as useful additional sources of reference:

Applied Clinical Trials (various issues)

Churchill's Illustrated Medical Dictionary (1989) New York: Churchill Livingstone.

Boyd KM, Higgs R and Pinching AJ (1997) *The New Dictionary of Medical Ethics.* London: British Medical Journal.

Bull K and Spiegelhalter DJ (1997) Survival analysis in observational studies. *Statistics in Medicine* **16**:1041–74.

Duncan AS, Dunstan GR and Welbourn RB (1981) *Dictionary of Medical Ethics*, revised edition. London: Darton, Longman and Todd.

Dupayrat J (1990) *Dictionary of Biomedical Acronyms and Abbreviations*, 2nd edition. Chichester: John Wiley & Sons.

Everitt BS (1995) *The Cambridge Dictionary of Statistics in the Medical Sciences.* Cambridge: Cambridge University Press.

Friedman LM, Furberg CD and DeMets DL (1985) *Fundamentals of Clinical Trials*, 2nd edition. Littleton: PSG Publishing Company.

Grieve AP (1998) *FAQs of Statistics in Clinical Trials.* Richmond: Brookwood Medical Publications.

Heister R (1989) *Dictionary of Abbreviations in Medical Sciences.* Berlin: Springer-Verlag.

Jadad A (1998) *Randomised Controlled Trials.* London: British Medical Journal.

Johnson FN and Johnson S (1977) *Clinical Trials.* Oxford: Blackwell Scientific Publications.

Last JM (1995) *A Dictionary of Epidemiology*, 3rd edition. New York: Oxford University Press.

Marriott FHC (1990) *A Dictionary of Statistical Terms*, 5th edition. Harlow: Longman Scientific and Technical.

Meinert CL (1986) *Clinical Trials: Design, Conduct and Analysis*. New York: Oxford University Press.

Meinert CL (1996) *Clinical Trials Dictionary: Terminology and Usage Recommendations*. Baltimore: The Johns Hopkins University.

Nahler G (1994) *Dictionary of Pharmaceutical Medicine*. New York: Springer-Verlag.

Pereira-Maxwell F (1998) *A–Z of Medical Statistics*. London: Arnold.

Po AL (1998) *Dictionary of Evidence Based Medicine*. Oxford: Radcliffe Medical Press.

Pocock SJ (1983) *Clinical Trials: A Practical Approach*. Chichester: John Wiley & Sons.

Rasch D, Tiku ML and Sumpf D (1994) *Elsevier's Dictionary of Biometry*. Amsterdam: Elsevier Science.

Raven A (1993) *Clinical Trials: An Introduction*. Oxford: Radcliffe Medical Press.

Samson P (1975) *Glossary of Bacteriological Terms*. London: Butterworth and Co (Publishers) Ltd.

Schwartz D, Flamant R and Lellouch J (1980) *Clinical Trials*. London: Academic Press.

Senn S (1997) *Statistical Issues in Drug Development*. Chichester: John Wiley & Sons.

Spilker B (1991) *Guide to Clinical Trials*. New York: Raven Press.

Spriet A and Simon P (1985) *Methodology of Clinical Drug Trials*. Basel: Karger.

Steen EB (1978) *Abbreviations in Medicine*, 4th edition. London: Baillière Tindall.

Vogt WP (1993) *Dictionary of Statistics and Methodology*. London: Sage Publications.

Winslade J and Hutchinson DR (1993) *Dictionary of Clinical Research*. Brookwood: Brookwood Medical Publications.

a posteriori   after the event; generally referring to decisions made or actions taken after data or results of a study have been seen. ⇔ *a priori*. ⇨ **Bayes' theorem, posterior distribution**

a priori   before the event; generally referring to decisions made or beliefs held before data or results of a study have been seen. Such decisions or beliefs may be based on data from previous studies or subjective feeling based on informal clinical experience. ⇔ *a posteriori*. ⇨ **Bayes' theorem, prior distribution**

Abbé plot   ≈ **L'Abbé plot**

abscissa   ≈ *x* **axis.** ⇔ **ordinate** (or *y* **axis**)

absolute change   the numerical difference between two numbers as in, for example, **change from baseline.** ⇔ **relative change**

absolute frequency   the number of items or the number of occurrences of a specified event. Often abbreviated simply to frequency. ⇔ **relative frequency**

absolute risk   the number of events (deaths, adverse reactions, etc.) divided by the number of individuals who could have experienced the event. ⇔ **relative risk**

absolute value   a numerical value that ignores any positive or negative sign; for example, the absolute value of $+3$ is $+3$; the absolute value of -3 is also $+3$

absorption   the process by which drug enters the blood stream. ⇔ **clearance, elimination**

absorption study   a study that measures the time taken for drug to be absorbed into the blood stream

accelerated failure time model   a statistical model used in **survival analysis** that assumes the effect of one treatment is to multiply the median **survival time** for patients randomised to one treatment group relative to that of patients randomised to another treatment group. ⇔ **Cox's proportional hazards model**

acceptance error   the error of accepting a statement (usually a **null**

hypothesis) when that statement or hypothesis is false. ⇔ **rejection error.** ⇨ **producer's risk, Type II error**

acceptance region   the values of a **test statistic** (for example, calculated values of *t* in a *t* **test** or of chi-squared in a **chi-squared test**) that lead to accepting the **null hypothesis.** ⇔ **rejection region.** ⇨ **critical value**

accountability   taking responsibility for one's own actions

accrue   to gather or accumulate (often with respect to patients, data or information)

accumulate   to collect more and more (patients, data, information, etc.) over time

accumulating data   when more and more data are available as time progresses. Usually used in the context of **sequential analysis** or **group sequential analysis**

accuracy   nearness of an **observed value** to its **true value** (even if the true value may never be known). Also used with respect to a measurement process to describe how closely that process measures the true quantity. ⇔ **precision**

accurate   close to the **true value.** ⇔ **precise**

active control   a **comparator group** in a study that receives an **active treatment.** ⇔ **placebo control**

active control equivalence study (ACES)   a study designed to show **therapeutic equivalence** between two active products

active ingredient   the pharmacologically or biologically active parts of product (the tablet, capsule, etc.) ⇨ **formulation, presentation**

active treatment   generally means a noninert pharmacological product or biological substance (not a **placebo**). The term is also sometimes used to describe the treatment of primary interest, rather than a comparator (but still active) treatment

actuarial method   ≈ **life table analysis**

acute   rapid onset and short lasting. A disease may be acute (for example chicken pox) as opposed to chronic (for example diabetes). Sometimes the term is used to describe part of a study that is used to treat the disease of interest, in contrast to a long term **follow-up period** looking for relapse or long term drug safety. Such a short term part of a study is sometimes called the acute phase of the study. ⇔ **chronic**

acute episode   short term appearance of symptoms of an underlying **chronic** (long lasting) illness. For example, bronchitis may be a chronic illness with acute episodes

acute phase   see **acute.** ⇔ **follow-up period**

acute study   short term study (usually of a long lasting disease).

⇔ **chronic study**

acute toxicity study   a study to investigate the short term **toxicity** of a product, usually a single dose of a drug. ⇔ **repeated dose toxicity study**. ⇨ **reproductive and developmental toxicity study**

ad hoc   one off. Something unique to a particular problem

adaptive design   study procedures that change as the study progresses. Most often refers to the details of the **randomisation** process changing as the study progresses and results become known. Such designs are used so that, if it appears that one treatment is emerging as superior to another, the **allocation ratio** can be biased in favour of the treatment that seems to be best. ⇨ **dynamic allocation**

adaptive inference   conclusions that can be made as data and information accumulate. Although this seems obvious, in many studies conclusions are drawn only once at the end of the study; adaptive inference may draw conclusions as the study progresses

adaptive randomisation   ≈ **adaptive design**

adaptive treatment assignment   ≈ **adaptive design**

additive model   a statistical model where the combined effect of separate variables contribute as the sum of each of their separate effects. ⇔ **multiplicative model**. ⇨ **interaction**

adequate and well controlled   a term describing a study that is sufficiently large, properly **randomised**, and **blinded**

adjust   to modify (usually the estimate of a **treatment effect**) to account for differences in patient characteristics between treatment groups. ⇨ **adjusted estimate**

adjusted estimate   an estimate of a **parameter** as would have been observed at some specified value of another variable. For example, high blood pressure (and its treatment) may be related to age and so we may wish to estimate the effect of a drug on people of different ages. ⇨ **analysis of covariance**

adjuvant therapy   extra treatment given to enhance the effect of a **monotherapy**. For example sensitising drugs to enhance the effect of radiotherapy

administer   to give (in the sense of giving treatment)

administrative review   a review of (usually accumulating) study data where the purpose is to monitor practical aspects of the study's progress (such as recruitment rates, shipment of laboratory samples, etc.) ⇔ **interim analysis**

admission criteria   ≈ **inclusion criteria**

adverse drug experience   ≈ **adverse event**

adverse drug reaction   ≈ **adverse reaction**

adverse drug reaction on-line information tracking (ADROIT)   a database kept of **adverse reactions** to marketed products

adverse event   any (usually) unwanted effect that a subject experiences whilst taking a drug. Note that **causality** is not implied. ⇔ **adverse reaction**

adverse experience   ≈ **adverse event**

adverse reaction   see **adverse event** but note that **causality** to a particular drug is implied

adverse treatment effect   ≈ **adverse reaction**

advocate   to support a given argument, opinion, or point of view

aetiology   the cause of a disease or the study of disease causality

agency   ≈ **regulatory authority**

aggregate   to combine separate data values into groups of **aggregate data**

aggregate data   data that have been grouped in categories. For example, all ages of patients in the range 0 to 5 put into one category, ages 6 to 12 in another category, etc.

agonist   a drug that enhances or activates the effect of a natural body chemical or of another drug. ⇔ **antagonist**

algorithm   a written description of a mathematical equation or **decision rule**. It is usually written partially in words (although not necessarily in complete and proper sentences) rather than just a set of mathematical expressions

all patients treated analysis   ≈ **intention-to-treat analysis**

all patients treated population   ≈ **intention-to-treat population**

all subsets regression   a method of deciding which variables should be in a **regression model**. ⇨ **backward elimination, forward selection**

allocate   to assign (typically a treatment to a patient) either by **randomisation** or by some deterministic method

allocation ratio   in a **parallel group study** the ratio of the number of patients allocated to one treatment group relative to the number allocated to another treatment group. Most often, the ratio is 1:1, or **equal allocation**

alpha (α)   the probability of making a **Type I error**. ⇔ beta (β). ⇨ **significance test**

alpha error   ≈ **Type I error**

alpha spending function   a method in **sequential studies** such that the times when **interim analyses** are performed do not need to be specified in advance. The number of, and timing of, interim analyses can be flexible

alphanumeric   data that may be alphabetical (a, b, c, . . . , A, B, C, . . . , including special symbols such as +, £, %) or numeric (0, 1, 2, . . . 9)

alternate allocation   a method of assigning treatments to patients whereby the first patient receives Treatment A, the second receives Treatment B, the third Treatment A, the fourth Treatment B and so on in a predictable (alternating) manner. ⇔ **random allocation**

alternative hypothesis (H_1)   this is usually the point of interest in a study. It is generally phrased in terms of the **null hypothesis** (of no treatment effect) not being true. If the objective of a study is to 'compare Drug A with placebo' then the null hypothesis would be that there is no difference between the two groups and the alternative hypothesis would be that there *is* a difference

alternative medicine   approaches to medicine such as homeopathy, acupuncture, herbal medicines, etc., considered by many people to be nonconventional medicines

altruism   putting the interests of the individual first; specifically in clinical trials, putting the interests of the individual before those of the research project. ⇨ **collective ethics, individual ethics**

amendment   ≈ **protocol amendment**

ampoule   ≈ **vial**

analysis   the process of summarising data or problems, describing them clearly (including plotting data) and drawing conclusions

analysis by administered treatment   a strategy where data are summarised and conclusions drawn based on the treatment that patients were actually given. ⇔ **analysis by randomised treatment**

analysis by assigned treatment   ≈ **analysis by randomised treatment**

analysis by randomised treatment   a strategy where data are summarised and conclusions drawn based on the treatment that patients were supposed to be given (the treatment they were randomised to receive), regardless of what they actually took. It is very similar to the term **intention-to-treat**. ⇔ **analysis by administered treatment**

analysis of covariance (ANCOVA)   a statistical analysis method that is an extension of **analysis of variance**. It allows estimates of **treatment effects** to be adjusted for possible **covariates** as well as **factors**

analysis of variance (ANOVA)   a statistical analysis method that allows comparison of two or more treatment groups and estimates of **treatment effects** to be adjusted for other possible **factors** such as race, gender, treatment centre, etc. It is a very general method covering a very broad range of techniques and can be used in a great variety of situations. Because of this, to describe a method of analysis as being 'analysis of variance' is rarely sufficient to adequately describe what analysis has actually been carried out

analysis policy $\approx$ **analysis strategy**

analysis population   the set (often subset) of patients recruited to a study who are subsequently included in the data analysis. Examples are the **all patients treated population, per protocol population**

analysis strategy   this combines the decision whether to use an **all patients treated analysis**, an **intention-to-treat analysis**, a **per protocol analysis**, or some other policy and considerations such as whether to use, for example, **parametric methods** or **nonparametric methods**, **Bayesian inference** or **frequentist inference**

anatomical therapeutic chemical classification system (ATC)   a drug **coding system** that codes according to a drug's site of action and its **indication**

and/or   a badly used term that often causes confusion, particularly over whether the word 'or' is considered as inclusive or exclusive. For example, if there are two events P and Q, one option is that both P and Q may occur; another option is that P or Q (but not both) may occur—this is the 'exclusive or'; finally P or Q (or both) may occur—this is the 'inclusive or'. If 'or' is considered inclusive then the term 'and/or' is redundant: 'P or Q' includes 'P and Q'; if 'or' is considered exclusive then the term may have some use. It is probably better to use a few more words and explain what is intended

anecdotal evidence   unsubstantiated evidence that cannot be strongly relied on. It is usually considered as more informed than mere opinion and often used as a means of generating ideas and research questions

aneugen   a substance that causes **toxic effects** on DNA. $\Rightarrow$ **clastogen**

angular transformation   a transformation applied to data that are of the form of **proportions to allow use of statistical methods based on the Normal distribution**. Where the proportion is p, the transformation is $y = \arcsin(\sqrt{p})$. $\Rightarrow$ **logistic function, probit transformation**

animal model   results from experiments in (nonhuman) animals, used to extrapolate results to humans

animal study   a study carried out in (nonhuman) animals. $\Rightarrow$ **preclinical study**

antagonist   a drug that prevents or reverses the effect of a natural body chemical or of another drug. $\Leftrightarrow$ **agonist**

antedependence model   a statistical method for analysing a series of **repeated measurements** on the same individuals. The method describes the data based (partly) on earlier measurements

applicable regulatory requirements   requirements of a **regulatory authority** that are either general to all studies or apply specifically to the experimental or geographical circumstances relevant to a particular study

approval   the process of an individual or group of individuals with appropriate authority agreeing to a request. This may take the form of approving a protocol, a submission to a research ethics committee, a submission to a regulatory authority, etc.

approximate   close to the **true value**. Note that the 'true' value may not be known and the interpretation of 'close' may vary from one situation to another, so this is a rather vague term

approximation   a method of estimating a **parameter** that gives an approximate answer

archive   to keep a historical record in secure conditions to confirm the data obtained and the procedures that were followed during the course of a study. ⇨ **backup**

arcsin transformation   ≈ **angular transformation**

area under the curve   a **summary measure** of data that have been collected repeatedly over time. The data are plotted with time on the *x* **axis** and the measurement on the *y* **axis**. The area is that between the line connecting the data points and the *x* axis (Figure 1)

arithmetic mean   ≈ **mean**

arm   synonym for group (as in **randomised group**)

artefact   an aspect of data that is not substantiated in other data sets and is not a real effect

ascending order   data sorted so that the smallest value comes first, the larger values later and the largest value last. This can be applied to alphanumeric data (by sorting into alphabetic order with special rules for including numbers and special symbols) as well as numeric data. ⇔ **descending order**. ⇨ **ranked data**

ascertainment bias   bias caused due to the manner in which data are collected. For example, surveying the general incidence of health problems near a doctor's surgery would probably lead to an unreasonably high proportion of respondents indicating less than perfect health; in contrast, surveying near a health club might lead to an unreasonably low proportion of respondents with impaired health

ASCII   a standard set of **alphanumeric** characters that is widely transferable between different computers. It stands for American Standard Code for Information Interchange

assay   a procedure to measure the quantity of a chemical (usually drug) in a sample (usually of blood or urine)

assent   agreement to something in a passive way and not after thorough consideration of the advantages and disadvantages. Note that clinical

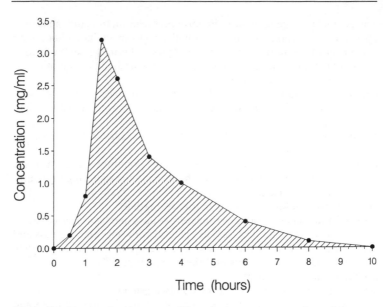

Figure 1   Area under the curve. Plot of serum concentration of drug on ten occasions up to 10 hours after administration. The area under the curve is shaded. Other features to note are C_{max} at 1.75 hours and $T_{max} = 3.2$ mg/ml

trials usually need subjects to **consent** to take part, not just assent

assessment   measurement of the state of disease. This may be a measurement of blood pressure, severity of depression, **quality of life**, etc.

assign $\approx$ **allocate**

assigned treatment   the treatment that a patient is due to receive based on a randomisation (or other) method

associate   an assistant (often in the sense of a **subinvestigator**)

associate investigator $\approx$ **subinvestigator**

association   a means by which two items are linked. For example, there is a link (or association) between smoking and lung cancer. $\Rightarrow$ **correlation**

assumption   a state (often a feature of data) that is taken as true although there may not be sufficient evidence to guarantee that state. A common assumption is that data come from a **Normal distribution**

asymmetric   not symmetric, as in not evenly split around the middle. The term is often used about **distributions** of data that are **skewed**

asymptote   a value that is never achieved but that is approached more and more closely. For example, repeatedly dividing a number by two will get closer and closer to zero but will never actually attain that value: in this case, zero is the asymptote

asymptotic method   a statistical method that assumes there is a large sample of data and which may not be suitable with small samples

atopy   indicates that an allergic disease such as asthma, eczema, etc. is hereditary rather than being a spontaneous new case

attenuation   making extreme results or statements less extreme and more typical of the norm

attributable risk $\approx$ **risk difference**

attribute   characteristic or feature (usually of a patient). All variables (age, sex, pulse, serum calcium, etc.) are attributes

attrition   loss; often used to describe loss of patients' data in **long term studies** due to patients withdrawing for reasons other than those of meeting the study's **primary endpoint**

audit   a systematic review of data and operational details or study procedures

audit certificate   a certificate to confirm that a study has been audited

audit report   a report (written or verbal) describing the findings of an **audit**. Such findings are usually restricted to points that do not meet expected standards of quality or completeness (rather than all aspects that do meet the expected standards)

audit trail   a list of reasons and justifications for all changes that are made to data or of all procedures that do not comply with agreed study procedures

auditor   a person responsible for carrying out an **audit**

autocorrelation   correlation between **repeated measurements** taken successively in time from the same subject

autoencoding   an automatic (usually by computer) method of assigning codes to data, for example codes for drugs or **adverse events**

autoregressive   a description of a process that produces data collected sequentially in time when each data point is potentially related (or **correlated**) with the previous one(s)

average   informal term for the **mean**

average absolute deviation   the average ($\approx$ **mean**) amount by which a set of data values differ from some **reference value** (usually that reference value being the mean). The differences ignore the sign (plus or minus). So, for example, the average absolute deviation of the numbers 1, 2 and

4 is $\{(2\frac{1}{3} - 1)+(2\frac{1}{3} - 2)+(4 - 2\frac{1}{3})\} \div 3 = 3\frac{1}{3} \div 3 = 1.11.$ ⇨ **standard deviation**

average deviation ≈ **average absolute deviation**

axis scale (*x* **axis** or *y* **axis**) on a graph

B

backup a reserve (often used in the sense of a reserve copy of data) kept under secure conditions in case of loss or corruption of the original. A more readily available and less permanent version of an **archive**

backward elimination a method of finding which variables should be kept in a **regression model** by including all possible variables and then removing ('eliminating') those that are deemed not useful. ⇔ **forward selection, all subsets regression**

backward stepwise regression ≈ **backward elimination**

bacterium single-celled microscopic organism; the cause of many diseases

Balaam's design a type of **crossover design** where patients are randomly assigned to receive treatments A and B in one of the **treatment sequences** AA, BB, AB or BA

balance the state of being equal, usually with reference to the number of subjects in each **treatment group**. ⇨ **balanced design**

balanced block part of an experiment (one **block** of it) such that within that block, the effect of each treatment is estimated with equal **precision**

balanced design an experiment in which the effect of each of the treatments is assessed with equal **precision**, usually by having the same number of subjects in each treatment group. Note that, in **crossover designs**, balance refers to there being as many **treatment sequences** AB as there are BA

balanced incomplete block design an experiment in which not all treatments being compared are represented in every **block** but where, overall, the occurrence of each treatment across all the blocks is the same (or **balanced**)

balanced randomisation a randomisation method which ensures that the effect of each treatment is estimated equally **precisely**, usually by assigning the same number of patients to each treatment group. ⇔ **unequal randomisation**

balanced study a study that has used **balanced randomisation**

bandit design ≈ **adaptive design**

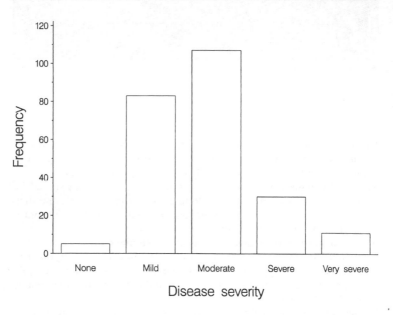

Figure 2   Bar chart. The number of patients who fall into each of the five categories is represented by the height of each bar

bar chart   a graphical method of showing the number of subjects that fall into each of two or more categories. The height of each 'bar' is proportional to the number of subjects within that category (Figure 2). ⇨ **histogram**

bar diagram   ≈ **bar chart**

Bartlett's test   a method of testing the **null hypothesis** that several **variances**, each estimated from different groups of subjects, are all equal

baseline   the moment in time that subjects are **randomised** or otherwise assigned their study medication. It is also used to refer to periods of time after a study has started but before randomisation has occurred

baseline characteristic   a measurement taken on a subject at the beginning of a study. Note that 'beginning' is generally taken to be at, or as near as possible to, the time of **randomisation**. ⇨ **demographic data**

baseline comparability   the process of, and results of, deciding if groups of patients assigned to different treatment groups (usually by **randomisation**) are similar with respect to **demographic data** and severity of disease

baseline data   ≈ **baseline characteristic**

baseline hazard function   in **survival analysis**, the **hazard function** for a subject in the **control group** (or a group arbitrarily chosen to be a control group)

baseline testing   see **baseline comparability**

baseline visit   usually the very first **visit** that subjects attend in a study. If **randomisation** does not occur at visit 1 then baseline visit may be used to refer to any visit before (and including) the **randomisation visit**

BASIC   a computer programming language. ⇨ **C**, **C++**, **Fortran**, **Visual Basic**

Baskerville design   a method for finding the most preferred of several treatments. Each subject is **randomly** assigned to a sequence of treatments but the length of time each patient receives each treatment is dependent on their own personal choice. If a subject is completely satisfied with the first treatment they receive then they would not change and would not receive any of the other treatments. In contrast, if a patient is not happy with any of the treatments being compared they would quickly pass through the entire possible set of treatments and finish the study

batch process   to work on a large number of documents all at once, rather than to handle each document as it arrives. This is a common term in data management but applies to computerised systems as well as manual systems

batch validation   to **validate** a large number of documents (usually data) as a **batch process**

baud rate   the speed at which data are transmitted electronically, measured as the number of binary digits sent per second. A baud rate of 32 000 means 32 000 binary digits sent per second

Bayes factor   the ratio of the **posterior belief** to the **prior belief**. This can be seen as a measure of how the **strength of evidence** in favour of a given **hypothesis** has increased, given new data, relative to the prior belief. ⇨ **Bayes' theorem**

Bayes' rule   the action one takes that gives the maximum **utility**

Bayes' theorem   the process of making judgements about the outcome of a study before the data are analysed (assigning **prior beliefs**), then combining these with the observed data (in the form of the **likelihood**) to obtain new **posterior beliefs**

Bayesian   general statistical methods based around **Bayes' theorem**

Bayesian inference   a method of **statistical inference** based on **Bayes'**

theorem as opposed to being based on **classical statistical inference** or **frequentist inference**

before–after design   a study in which subjects are observed before treatment is given and their disease state and severity is recorded. These subjects are then given treatment and subsequently their disease state and severity are reassessed. ⇨ **crossover design**

Behrens–Fisher problem   the problem of using a statistical **significance test** to compare two **means** when their **variances** are not equal. Behrens and Fisher originally discussed the problem. Note that the usual *t* **test** assumes that the variances in the two groups are equal. It is a long standing mathematical and philosophical issue; hence being referred to as the Behrens–Fisher 'problem' rather than the Behrens–Fisher 'method'

bell shaped   used to describe a **distribution** that when drawn as a **histogram** or **density function**, has the same profile as a bell. The **Normal distribution** (see Figure 22) is the most common example but the term should not be used exclusively for that purpose

benchmarking   the process of comparing activities (usually performance measures) against a standard reference value or in the absence of a standard, then against other methods to achieve the same outcome. Examples commonly include the costs of running studies between different companies; speed of recruitment into studies in different therapeutic areas, etc.

beneficial effect   a **therapeutic effect** of a drug that is considered to be advantageous to the patient. It is usually meant to imply alleviating symptoms of the disease under study but is not limited to that. If a topical treatment were intended to alleviate symptoms of rash on the scalp and it appeared to reverse the effects of alopecia then the effect on alopecia would be considered a beneficial effect. ⇔ **adverse event**

benefit   a nontechnical term referring to advantage (of one treatment or activity over another). It may be measured in a variety of ways including decreased cost, increased patient satisfaction, reduced length of hospital inpatient stay, extended life expectancy

benign   a condition that does not produce any harmful effects

Berkson's fallacy   drawing wrong conclusions (usually in **case-control studies**) because of **selection bias**

Bernoulli distribution   the **probability distribution** of a **binary variable**

best case analysis   the process of making assumptions, often about data that are missing (either inadvertently or because they could not be measured), when the implications of those assumptions are that a treatment may appear to give more benefit than is truly justified.

⇔ **worst case analysis.** ⇨ **sensitivity analysis**

best fit   used in the context of **regression** and fitting lines (straight or curved) to data on graphs. The best fitting line is generally the one that has the data points closest to the **regression line** (although various other criteria for 'best' may be specified)

best linear unbiased estimator   a **linear estimator** that is better (usually in the sense of having a smaller **variance**) than any other possible linear estimator

beta (β)   the probability of making a **Type II error.** ⇔ **alpha (α).** ⇨ **significance test**

beta coefficient   ≈ **regression coefficient**

beta error   ≈ **Type II error**

beta level   the probability of making a **Type II error**

between groups   usually used in the sense of estimating the variation (strictly speaking the **variance**) of data where we are describing the variation between the means of two or more groups of subjects. ⇔ **between subjects, within groups**

between groups sum of squares   a measure of **variability** (by the method of **sum of squares**) between different groups (treatment groups, strata, etc.) in a study. ⇔ **within groups sum of squares**

between groups variance   ≈ **between groups**

between groups variation   an informal term for the **between groups variance**

between person   ≈ **between subjects**

between study   in **meta-analyses** this is used to describe the variation that is due to differences between studies rather than differences between subjects within each study

between study variance   see **between study**. This makes it clearer that it is the **variance** (or variation) that is being considered

between study variation   a less formal term for **between study variance**

between subjects   usually used in the sense of estimating the variation (strictly speaking the **variance**) of data where we are describing the variation between individual subjects. ⇔ **between groups, within subjects**

between subjects comparison   the types of analyses that are made in **parallel group studies**, that are unpaired comparisons, rather than **paired comparisons**

between subjects effect   ≈ **between subjects comparison**

between subjects study   ≈ **parallel group study**

between subjects sum of squares   a measure of **variability** (by the method of **sum of squares**) between different subjects in a study. ⇔ **within subjects sum of squares**

between subjects variance   see **between subjects**

between subjects variation   an informal term for the **between subjects variance**

between treatments $\approx$ **between groups**

bias   a process which systematically overestimates or underestimates a **parameter**. Bias is sometimes, but not always, acceptable: for example, we routinely underestimate peoples' ages by an average of 6 months if we record data only to the lowest whole year. $\Rightarrow$ **precision**

biased coin   a method of **randomisation** that does not assign patients to treatments with equal probabilities. $\Leftrightarrow$ **balanced design.** $\Rightarrow$ **unequal allocation**

biased estimator   a method of **estimation** of a **parameter** from data that gives a biased result

bibliography   a list of published books, manuscripts, etc. that discuss a particular topic

bimodal   having two **modes**

bimodal distribution   a distribution (either a **probability distribution** or a **frequency distribution**) that has two **modes** or **peaks**

binary data   data taking only one of two values: typical examples are data of the form Yes/No, Dead/Alive, Male/Female. Sometimes a third category of 'not known' or 'missing' is included but the data are still said to be binary. $\Rightarrow$ **categorical data**

binary outcome   an **outcome** that can take only one of two values; one that yields **binary data**

binary variable   a **variable** that can take only one of two values; one that yields **binary data**

binomial data $\approx$ **binary data**

binomial distribution   in data that are binary (yielding only 'positive' or 'negative' outcomes), the **probability distribution** of the number of positives is a binomial distribution. For example, the number of live births (as opposed to still births) out of the first one hundred deliveries in a maternity unit follows a binomial distribution

bioassay   estimation of the **potency** of a drug by observing its effect on a biological organism

bioavailability   at any time, the proportion of drug within the body that is available to give a **therapeutic effect**

biochemistry   the study of chemistry in living things. Usually used in the context of **laboratory data** to refer to the amount of various chemicals (for example albumin, calcium, ethanol) in the blood. $\Leftrightarrow$ **haematology**

bioequivalent   two products that have the same **bioavailability** are said to be bioequivalent

biologic   a drug derived from a biological product. ⇨ **biotechnology**. ⇔ **pharmaceutical, phytomedicine**

biological assay   ≈ **bioassay**

biological marker   a **nonclinical** (often a laboratory) measurement that is an indicator (or 'marker') of a clinical condition

biological plausibility   a **hypothesis** that is justifiable from biological theory and not just based on observable data

biometrician   a person who specialises in biological (including medical, genetic, agricultural) applications of **statistics**

biometry   literally 'measurement in biology'. More generally, the application of statistical theory and methods in the biological sciences

biopharmaceutical   the subset of biology related to **pharmacology**. Often the term is used synonymously with **pharmaceutical**

biostatistician   ≈ **biometrician**

biostatistics   the application of statistical theory and methods in the biological sciences

biotechnology   the process of developing drugs from biological products (such drugs are then called **biologics**)

bivariate   the joint measurement and consideration of two characteristics (for example a person's 'size' would often be measured in terms of their height and weight). ⇨ **multivariate**

bivariate analysis   special methods of analysis suitable for **bivariate data**. These are usually simplifications of general methods of **multivariate analysis**

bivariate data   measurements that consist of two **response variables**. For example, a person's blood pressure could be measured as both systolic pressure and diastolic pressure. More than two variables are always referred to as **multivariate data**

bivariate distribution   the **joint distribution** of two separate (but often **correlated** or related) measurements. ⇔ **univariate distribution**. ⇨ **multivariate distribution**

black box   a process whose internal workings are unknown (at least to the user) but whose output is usually trusted. Computers, for example, are black boxes to most people

blind   not being able to see. Specifically, within clinical trials, where the investigator, subject (and possibly other people) are not able to distinguish different treatments that are being compared (by sight, smell, taste, weight, etc.) ⇨ **single blind, double blind, triple blind, quadruple blind**

blinding   the process of keeping hidden certain information about data or study procedures in order to help avoid **bias**. Most commonly this means keeping the **treatment allocation** hidden from the doctors and patients (and often data management staff) taking part in a study

block   several packs of medication kept together and used **sequentially**, each block usually having the same number of **treatments** (although in **random order**) as each other block. The concept can be extended to cases when treatment 'packs' do not actually exist. Commonly, if a study is comparing two treatments each 'block' might contain medication for four patients, two on one treatment and two on the other. ⇨ **block size**

block effect   any **systematic** difference in response that may exist between blocks of treatment medication. Such differences do not invalidate the study; the purpose of blocking is to ensure that if such differences exist, the **treatment allocation** is equal across blocks

block size   the number packs of **treatment** that form one complete **block**

blocked randomisation   a randomisation scheme that uses **blocks** to help maintain **balance**. ⇔ **completely randomised design**

blocking   the act of using **blocks** of treatment

body-mass index   ≈ **Quetelet's index**

Bonferroni correction   an adjustment made when interpreting **multiple significance tests** that all address a similar basic question. If two endpoints have been assessed separately, instead of considering whether a *P*-**value** is less than (or greater than) 0.05, the calculated *P*-value should be compared to 0.025. In general, if k *P*-values have been calculated, the declaration of **statistical significance** should not be made unless one or more of those *P*-values is less than $0.05/k$

Boolean logic   rules for making decisions based on combining **binary outcomes** using the key words AND, OR and NOT. For example, subjects may be eligible for a study 'IF (they are male) OR ((they are female) AND (they are using adequate contraception))'

bootstrap   a **simulation** method used for statistical **significance testing** and **estimation** that takes as possible (simulated) sample data values, only those data values that have actually been observed. ⇨ **Monte Carlo method**

box and whisker plot   a diagram used to show a few key features of a **frequency distribution**, namely the minimum, **lower quartile**, **median**, **upper quartile** and maximum (Figure 3). ⇨ **Exploratory Data Analysis**

box plot   ≈ **box and whisker plot**

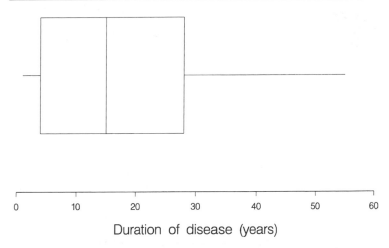

Duration of disease (years)

Figure 3   Box and whisker plot. Distribution of the number of years a group of 66 patients had suffered from eczema. The key features illustrated are the minimum (1 year), the lower quartile (4 years), the median (15 years), the upper quartile (28 years) and the maximum (55 years)

Box–Cox transformation   a very general equation used to transform a set of data so that it better resembles a **Normal distribution**. The method was developed by the two statisticians, Box and Cox; hence the name

branch   on a **decision tree** (Figure 6), any of the possible routes that can be followed. ⇨ **node**

brand name   ≈ **trade name**

break point   used to describe a **regression line** that is not a continuous smooth function across the whole range of data but is made up of different lines (often only two). The point at which the two lines (with different slopes) meet is called the break point

bridging study   a study designed to extend the applicability of a **confirmatory study**, usually to broaden the population to which the results apply. Bridging studies are usually much smaller than other confirmatory studies

byte   a single character (one letter or a single **digit** in a number) as stored by a computer

C a widely used, high level, computer programming language. There are other programming languages that are commonly used in clinical trials work such as **BASIC, C++, Fortran, Visual Basic** and there are specific statistical analysis programs, for example BMDP®, SAS®, SPSS®, STATA®

C++ a more advanced version of the **C** programming language

calibrate to check measurements against a known standard

capsule a dissolvable container (with an **enteric coating**) that contains a drug. ⇨ other **delivery devices** such as **tablet, transdermal patch**

carcinogen a chemical that causes any type of cancer

carcinogenicity the potential to cause any type of cancer

carcinogenicity study a study to determine if a chemical is a **carcinogen**

carryover a term used mostly in the context of **crossover studies** where the effect of a drug is still present after that drug has ceased to be given to a subject, and in particular when that subject is taking another drug

Cartesian coordinate the place (in terms of x axis and y axis) where a data point lies on a graph. For example, if a subject's systolic blood pressure is 120 mmHg and their diastolic blood pressure is 85 mmHg the Cartesian coordinates would be 120, 85

case a term used synonymously with **patient**, although often intended to mean one with a particular identified disease. ⇔ **control**. ⇨ **case-control study**

case history the description (usually the **medical history**) of an individual case

case record form (CRF) the term used for the paper on which data are written. Often a CRF comes in the form of a book with many pages of forms to record a subject's data

case report form (CRF) ≈ **case record form**

case-control study a type of study used for evaluating the causes of a particular disease. A group of patients with the disease (the **cases**) are compared with another group of subjects who do not have the disease (the **controls**). Their lifestyles, previous exposure to potential hazards, demographics, etc., are compared to try to distinguish which of those features predispose someone to have the disease in question

case-fatality rate   the **death rate** amongst a group of **cases**

catchment area   the geographical area from which subjects may be included in a study. The area covered by a Health Authority or in which a survey was being carried out would be termed the 'catchment area'

categorical data   data that are not pure measurements but are in the form of labels assigned, such as 'male' and 'female'. ⇨ **ordered categorical data**

categorical scale   the scale on which **categorical data** are measured

categorical variable   a characteristic of a subject that results in **categorical data**. For example a subject's gender is a categorical variable: it falls into the categories 'male' or 'female'

categorise   the process of taking data that may take many distinct values and putting them into categories. For example all the people living in a group of post codes may be classed (or categorised) as being in one town

category   a **group** (but used when the term is applied to data). Categories of blood group are A, AB, O; categories of products might be 'prescription only' or 'over the counter'

causal relationship   a relationship that is observed when one variable is a consequence of another. For example, alcohol intake and impaired reaction times are causally related. ⇨ **correlation**

causality   the act of causing. ⇨ **correlation**

cause and effect   a phrase used to imply **causality**, over and above **correlation**

ceiling effect   a term to describe an **asymptote** that is an upper limit. ⇔ **floor effect**

cell   when referring to a tabulation of data, each of the individual categories or subcategories of patients (or other data) are referred to as cells

cell frequency   the number of subjects within a **cell** of a table. For an example, see **contingency table**

cell mean   the mean of the data for all subjects within a **cell** of a table (Table 1)

censor   to prevent something being observed. ⇨ **censored data**

censored data   when the time until an event (typically cure, recurrence of symptoms or death) is the data value to be recorded and that event has not yet been observed for a particular subject, that data value is said to be censored. ⇨ **truncated data**

censored observation   ≈ **censored data**

centile   if a (large) set of observations are placed in order, the 1st centile is the value below which 1% of the data lie, the 2nd centile the value below

Table 1 Cross-classification of mean systolic blood pressure (mmHg). The data are cross-classified by treatment group and by centre. Each of the means (127.4 mmHg, 135.1 mmHg, etc.) are referred to as cell means

Centre	Treatment A	Treatment B
UK.001	127.4	135.1
UK.002	131.2	135.9
UK.003	122.0	127.6
UK.004	129.5	130.3
UK.005	141.3	147.7

which 2% of the data lie, etc. ⇨ **lower quartile, median, upper quartile**

central laboratory   a single laboratory that is used by all centres in a **multicentre study**, though it may not necessarily be 'central' in any geographical sense. ⇔ **local laboratory**

central limit theorem   a statistical phenomenon such that the mean of several data values tends to follow a **Normal distribution**, even if the distribution of the original data was not Normal

central processing unit (CPU)   the part of a computer that carries out calculations

central randomisation   in **multicentre studies** it is common to use a separate **randomisation list** in each centre so that we **stratify** the randomisation by centre. Alternatively we may have a single randomisation sequence, held at one site (a 'central' site) and investigators would telephone (or otherwise contact that site) to obtain the next randomisation code

central range   the range in which the central 90% of the data from a distribution lie. ⇨ **interquartile range, standard deviation**

central tendency   a nonspecific summary of data (usually of **continuous data**) that, for any particular purpose, is useful in describing where the bulk of the data lie. The **mean, median** and **mode** are the most common measures of central tendency

certificate of destruction   an officially recognised document confirming that specific batches of drug product have been safely destroyed. This would often apply to unused medication from a study

certified   officially recognised (strictly speaking, with a certificate). This can refer to an individual, a machine, a blood sample, etc.

challenge test   the administration of a product specifically to see if it produces an **adverse reaction**

chance   luck (good or bad). Events that happen by chance are ones that

Table 2   Individual subjects' systolic blood pressure (mmHg) before and after treatment

Subject identification number	Before treatment (baseline)	After treatment	Change from baseline
1	137	134	−3
2	120	120	0
3	150	163	13
4	118	126	8
5	130	135	5
6	130	122	−8

could not have been predicted with certainty. They occur with probability less than one

change from baseline   when a measurement (for example subjects' blood pressure) is measured at the time of randomisation (the **baseline**) and again after treatment and the difference calculated (as in Table 2), this difference is often used as the measure of treatment benefit. It is called the 'change from baseline'. ⇨ **analysis of covariance**

change score   ≈ **change from baseline**

changeover design   ≈ **crossover design**

changepoint model   a statistical model that attempts to identify when a smooth course of events abruptly changes. For example, height of growing children may follow a smooth curve until puberty, when a sudden change in that curve would be expected. A model that allows for this would be called a changepoint model

characteristic   an alternative term for data or measurement. Often (but not necessarily) restricted to **demographic data** and **baseline data**

chart   a general term for any form of graph, histogram, etc.

check   to confirm that something (often data) is correct

check digit   a number (usually between 0 and 9) that is used as a means of checking that other numbers are correct. The last digit of the ISBN of this book is the check digit

chemotherapy   the use of drugs to eradicate disease or to prevent existing disease from spreading by killing cells that are otherwise dividing and multiplying. ⇨ **cytotoxic**

chi-squared (χ^2)   ≈ **chi-squared statistic**

chi-squared distribution   a **probability distribution** used in a wide variety of forms of data analysis. Most often in clinical trials it is used for comparing the equality of proportions in **contingency tables**. However, its use is not restricted to this case

chi-squared goodness of fit test   ≈ **chi-squared test**

chi-squared statistic   the calculated value of **chi-squared** from a set of data

chi-squared test   a statistical **significance test**, the most simple of applications being for testing the **null hypothesis** that two (or more) proportions are equal

chronic   long term. ⇔ **acute**

chronic study   a study of the long term treatment of a disease. ⇔ **acute study**

chronobiology   the study of how biological features change with time. Biological example of **time series** methods

chronotrophic effect   the effect of a drug on the force of the heart beating. ⇨ **inotropic effect**

CIOMS form   a standard template form for reporting **adverse events** to **regulatory authorities**. CIOMS stands for Council for International Organisations of Medical Sciences

circadian rhythm   a biological process that repeats itself in 24-hour cycles

citation   the reference in a published paper, book, etc., to another previously published piece of work

class   ≈ **category**

class interval   when **continuous data** are categorised into groups (for example, age groups) the class intervals are the number of years grouped into each category. They may, for example, be ten-year age groups of 0–9, 10–19, 20–29, etc. It is a term generally used when all classes have the same interval but this need not necessarily be the case

class limits   when **continuous data** (for example age) are categorised into groups (age groups) the class limits are the values that define at what values each group starts and finishes. The age groups may be 0–15 years, 16–64 years and 65–75 years; these values would be the class limits. Note that there is no need (explicitly or implicitly) for the **class interval** to be the same for every class

classical statistical inference   statistical methods that rely heavily on **significance testing** and calculating **confidence intervals**. ⇔ **Bayesian inference**

classification variable   a variable that is used to assign patients into groups (for example blood group, ethnic origin, or a **continuous variable** such as blood pressure that has been categorised)

classify   to assign a subject to a group (or category), based on data

clastogen   a substance causing damage to genetic material. ⇨ **aneugen**

clean data   data that contain no errors. Often data that are believed to contain no errors are referred to as 'clean' but there is an assumption that may not be valid. ⇔ **dirty data**

clearance   the rate of **elimination** of a drug from the body as a proportion of the amount of drug in the body. ⇔ **absorption**

clinic   a medical centre where people are cared for

clinical   the branch of medicine dealing with **patients**. The practical application of medicine rather than medicine as a pure subject. ⇔ **medical**

clinical ethics   ethical considerations and behaviour concerned with treating an individual subject. ⇔ **research ethics**

clinical investigation   any form of investigating a patient or analysing a sample from a patient to help determine a **diagnosis**

clinical investigational brochure   a document describing the full extent of knowledge concerning an **investigational product**

clinical practice   what is generally accepted as the way patients are cared for. This includes all aspects of patient care including waiting in outpatient clinics, drugs received, palliative care, etc.

clinical research   that area of research carried out on humans (either patients or healthy volunteers). ⇔ **preclinical research**

clinical research associate   someone employed to monitor the organisation and practical issues to do with running a study. Their duties may include collecting and collating study documentation, ensuring complete and clean data, ensuring that pharmacies or other dispensing centres have adequate supplies of materials

clinical research coordinator   ≈ **clinical research associate**

clinical research organisation   a company that provides staff and/or facilities for carrying out clinical studies

clinical significance   a finding or observation that is **clinically significant** (for example a patient dying unexpectedly or a large treatment effect). ⇨ **clinically significant difference**. ⇔ **statistically significant**

clinical study   any systematic study that includes **patients**. This need not include studying any **treatments**, for which ≈ **clinical trial**

clinical trial   any systematic study of the effects of a treatment in human subjects. ⇨ **Phase I, Phase II, Phase III, Phase IV**. Note that although **randomisation** and **blinding**, for example, are considered as some of the essential features of good clinical trials, these are not requirements. ⇨ **prevention study**

clinical trial certificate (CTC)   the certificate issued before the introduction of the system of the **clinical trial exemption certificate (CTX)**. The amount of information needed for a CTC is more than that required for a CTX

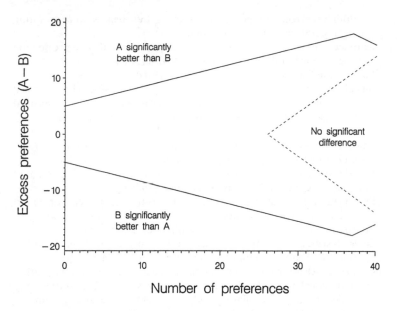

Figure 4 Closed sequential design. The solid lines indicate stopping boundaries for declaring a statistically significant difference between treatments A and B. For example, if out of ten patients expressing a preference for one or other treatment, nine preferred treatment B and only one preferred A, then the study would stop, concluding that B is significantly better than A. If the broken boundary is crossed, then the study stops and the conclusion is drawn that no significant difference was found between the treatments

clinical trial exemption certificate (CTX)   a certificate (issued by a **regulatory authority**) to a pharmaceutical company authorising use of an unlicensed product or use of a product outside its **marketing authorisation** for the purpose of carrying out a clinical study. Note that it is the product that is being exempted from otherwise stringent rules, not the study, so that one CTX may serve to cover several studies.
   ⇨ **doctors and dentists exemption (DDX)**
clinically important   ≈ **clinically significant**
clinically meaningful difference   ≈ **clinically significant difference**

clinically significant   an effect (in an individual subject or an average effect in a group of subjects) that is sufficiently large to be of benefit (or harm) to a patient or of note to a treating physician

clinically significant difference   a treatment effect that is sufficiently large to be useful for treating patients

closed sequential design   a **sequential study** design that does not have a predetermined number of patients. An upper limit on the number of patients does exist (hence 'closed') but it is possible to draw conclusions and stop the study before that number of patients has been recruited (Figure 4). ⇔ **open sequential design**

closed sequential study   a study that is designed as a **closed sequential design**

cluster randomisation   a case where individual subjects are not **randomised** to receive different interventions, but rather groups ('clusters') of subjects are randomised. Examples are most common in **community intervention studies** where, for example, some towns may have fluoride introduced to their water supply whilst other towns may not. Clearly each member of the community cannot be randomly assigned to have fluoride, or not, and the randomisation must be done in large groups of subjects (or clusters)

C_{max} the maximum concentration of drug measured (usually) in a subject's blood but it could also apply to that measured in urine. The term can also be used to refer to the mean of many subjects' C_{max} values; it is then used as a description of the product rather than of any particular subject. (≈ Figure 1, **area under the curve**) ⇨ T_{max}

coarse data   data that are measured or subsequently recorded very approximately, for example in categories with large **class intervals**. ⇔ **fine data**

code   an indirect means of linking two or more pieces of information. For example, to identify a pack of medication that pack may be given a code number and, separately, a list be kept of which code numbers refer to which treatments. ⇨ **randomisation code**

coding dictionary   a list of terms and associated codes. See, for example, **COSTART, MedDRA, WHO-ART**

coding system   a set of rules for making up **codes** for data

coefficient   an estimate of a **parameter**. The term is used when the parameters are being estimated in statistical models such as **regression analysis, logistic regression**

coefficient of concordance   a measure of agreement between several people, each rating a group of items on some specific measure. ⇨ **correlation**

coefficient of determination   the square of the **correlation** between two variables, denoted r^2

coefficient of variation   a measure of variation in data, relative to the mean of those data. It is calculated as $100 \times$ (standard deviation/mean) and is expressed as a percentage

cohort   a group of individuals with a common characteristic observed over a period of time. The feature they have in common may simply be the year of birth, or it may be the fact that they have all been exposed (for example) to a **carcinogen** or a novel educational programme

cohort effect   any systematic difference between subjects recruited to a study at different times. For example, the first patients recruited to a study may have less (or possibly more) severe symptoms than those recruited later in the study

cohort study   the study of a group of subjects over time. This includes clinical trials, but the term is usually restricted to observational studies

co-intervention   more than one **intervention** being studied concurrently. Note that the interventions do not necessarily have to be given at the same moment but the period of study is coincidental, nor do both interventions need to be related to the same disease or be of the same type. For example, a drug treatment and a patient management strategy might both be studied concurrently

collapse   used in the sense of reducing the number of **categories** of data. For example, age may be recorded as under 5 years, 5–15 years, 16–65 years, etc. Subsequently deciding to combine adjacent categories (for example the under 5s and the 5–15s) would be described as 'collapsing' these two categories into one

collective ethics   ethical behaviour that is more concerned with benefiting other people than oneself. Being prepared to administer a placebo is unlikely to benefit the patient concerned but may benefit others by nature of the information gained. ⇔ **individual ethics**

column vector   see **vector**

combination drug   more than one drug being administered simultaneously (usually when all of the drugs are packaged in the same tablet or capsule, etc.) ⇔ **monotherapy**

community intervention study   a study carried out to investigate the effect of an **intervention** on an entire group of people, for example all those who live in a particular city. Public health studies and studies of **screening programmes** frequently are described as community intervention studies. ⇨ **cluster randomisation**

community study   a study of large numbers of subjects in a community. It could be some kind of survey or might be a **community intervention study**

comparability   similarity. Often used in the sense of describing how similar two randomised groups are with respect to demographic data or disease severity

comparable   the state of being similar

comparator drug   ≈ **comparator treatment**

comparator group   the group of patients assigned to receive the **comparator treatment**

comparator study   a study that makes comparisons (usually between treatments). ⇔ **observational study**

comparator treatment   usually the drug, placebo, or other intervention with which a new or **experimental treatment** is being compared

comparison   a **contrast** (formal or informal) between two or more items or groups

comparison group   ≈ **comparator group**

comparisonwise error rate   the probability of making a **Type I error** for each comparison in a study. ⇔ **experimentwise error rate.** ⇨ **multiple comparisons**

compassionate use   a regulatory term, meaning that an unlicensed product is allowed to be used for a limited number of patients for whom their is no alternative medication. Although the product may be ineffective (its efficacy has not been demonstrated), there may be no other effective therapies. ⇨ **named patient use**

compassionate use protocol   a **protocol** that defines how a product will be used on a **compassionate use** basis

competitive enrolment   the situation in **multicentre studies** where each centre is allowed to recruit as many subjects as they can until the overall **recruitment target** has been met, rather than each centre having their own recruitment target

complementary log – log transformation   an equation applied to data that are proportions to allow use of statistical methods based on the Normal distribution. The transformation is $y = \log(-\log(1-p))$

complete block   a **block** of medication that contains all possible treatments (or combinations of treatment or **treatment sequences**) that are being studied. ⇔ **incomplete block**

complete block design   a study design that only uses **complete blocks** of treatment. ⇔ **incomplete block design**

complete cases analysis   a strategy for analysing data where only subjects who provide complete data are included in the analysis; any subject with **missing data** is excluded. ⇨ **intention-to-treat**, **per protocol analysis**

complete response   in cancer studies, this is generally regarded as complete disappearance of all tumours and no new tumours. ⟹ **partial response, stable disease, progression**

completely randomised design   a study where subjects are allocated to receive treatments in a randomised manner with no constraints (such as equal numbers of patients per group, no **blocks**, no **stratification**)

compliance   the measurement of how fully patients take their medication. This may be measured by weighing returned medication, counting returned tablets or simply asking how many doses of medication were (or were not) taken

compliant   ≈ **fully compliant**

component   a part. This may be a chemical component of a drug (one of the chemicals that make it up), a part of a data file or of a case record form, etc.

components of variance   a method of analysing data that assesses which features of an experiment account for the variation in those data. Typically, the sorts of features identified will be patients, treatment centres and different medications

composite hypothesis   in a statistical **significance test**, an **alternative hypothesis** that does not specify a single value for a **parameter**, for example $H_1{:}\mu > 0$ ⟺ **simple hypothesis**

composite outcome   when an **outcome measure** for a study is a mix of several individual measurements. For example, the composite outcome 'treatment success' may be defined as a patient who is free of symptoms and has a quality of life score better than some specified value. Neither of those features is sufficient on their own to define a treatment success but together they are. ⟹ **Guttman scale**

composite score   ≈ **Guttman scale**

compound   the bulk product of drug. ⟺ **product**

compound symmetry   a term used in assessing **repeated measurements**. The data are required to have the same **variance** at each time point and equal **covariances** between time points. Generally, if compound symmetry can be assumed, the analysis of data is much simpler

computer   a machine (originally mechanical but now electronic) used for numerical calculations and data processing. The current uses range from complex and fast calculations through to controlling machinery and word processing

computer assisted data collection   a process by which a computer is used to help (in various possible ways) collection and/or recording of data. The help may simply be that it acts in the form of an electronic **case**

record form and that data are recorded into the computer instead of onto paper. It may be more sophisticated and the computer linked to a holter monitor to directly record measurements of blood pressure without the need for human intervention

computer assisted new drug application (CANDA)   a **new drug application** where some or all of the data, study report, program files, etc. are supplied to the **regulatory authority** in electronic form on a computer

computer package   a **computer program** that does a variety of related tasks

computer program   instructions given to control what a computer does. A variety of types of computer programs are used in clinical research including those for data processing, statistical analysis, drawing graphics and report writing

concentration   the amount of a substance in a fixed volume of liquid. This may be the amount of active drug per unit of blood during **absorption** and **distribution**

conclusion   the decision that is made based on data that have been collected and analysed. Note that results should generally be referred to in the past tense ('Drug A was better than Drug B') but conclusions should be referred to in the present tense, with future implications ('we conclude that Drug A is better than Drug B'). ⇔ **discussion**

concomitant medication   drugs that are not being studied but which a patient is taking through all or part of a study. These may be other drugs for the same indication as the study or for other indications

concomitant variable   a variable that may influence the results of a study but which is not a part of the study design. Most often, this term is used to refer to other (nonstudy) medications that a patient may be taking or other diseases that a patient may have. ⇨ **concomitant medication**, **covariate**

concordance   agreement. ⇨ **coefficient of concordance**

concordant pair   in a study where subjects are assessed on two different occasions or by two different measuring devices and the variable measured is **binary** (for example, disease present or absent), the data may be summarised in a **two-by-two table**. The concordant pairs are those pairs of observations where the two measurements agree with each other. ⇔ **discordant pair**

concurrent control   control subjects who are observed and data recorded concurrently with the active subjects. This need not necessarily be done in a controlled experiment. ⇔ **historical control**

concurrent medication   ≈ **concomitant medication**

conditional distribution   the **distribution** of one variable at a fixed value of another variable. For example, the distribution of age may be given for all subjects, but if it is given for males and females separately then these sex-specific distributions are said to be 'conditional on sex'. ⇔ **joint distribution, marginal distribution**

conditional odds   the **conditional distribution** of the **odds** of an event occurring

conditional power   the **power** of a study based on some prerequisite information. Usually it is meant as the power of the study as calculated (after the study has finished) using the **observed difference** between the treatments and the **observed variance** of that difference

conditional probability   the **probability** of an event happening, given that another event has already been observed to happen

confidence   an informal term used to describe how strong is one's belief in the results of a study. ⇨ **strength of evidence**

confidence coefficient   see **confidence interval**

confidence interval   a range of values for a **parameter** (such as a mean or a proportion) that are all consistent with the observed data. The width of such an interval can vary, depending on how confident we wish to be that the range quoted will truly encompass the value of the parameter. Usually '95% confidence intervals' are quoted. These intervals will, in 95% of repeated cases, include the true value of the parameter. In this case, the **confidence coefficient** (or **confidence level**) is said to be 95% (or 0.95). Confidence intervals are a preferred method of estimating parameters, whilst **significance tests** compare those parameters with arbitrary values. ⇨ **posterior distribution**

confidence level   see **confidence interval**

confidence limit   the values at the end of a **confidence interval**. If the 95% confidence interval for the difference in mean systolic blood pressure between two treatment groups is quoted as being from $-3\,mmHg$ to $+8\,mmHg$, then -3 and $+8$ are the confidence limits

confidential   private; not to be disclosed to a third party

confirmatory analysis   the analysis of a **confirmatory study**

confirmatory study   a study that is designed to answer a specific question without leaving any room for doubt. Whilst **Phase I studies** and **Phase II studies** give some information regarding efficacy and safety, **Phase III studies** are usually thought of as being confirmatory. ⇔ **exploratory study, pilot study**. ⇨ **definitive study**

conflict of interest   the situation where an individual or organisation may find it difficult to make unbiased statements. Examples are of investigators reviewing their own project proposals at an ethics committee meeting or a pharmaceutical company reporting results of a study involving one of their own products. In such cases, bias is not being assumed but it is recognised that there is a clear reason why individuals may make biased statements or give biased opinions

confounded   'cannot be distinguished from'. For example, if all males were given one treatment and all females given an alternative, the effects of treatment and gender would be indistinguishable from one another, or confounded with each other

confounder   a term used in **observational studies** to describe a **covariate** that is related to the outcome measure and to a possible **prognostic factor**

confounding factor $\approx$ **confounder**

consent   positive agreement, particularly in the sense of **informed consent**. $\Leftrightarrow$ **assent**

conservative   erring towards being safe; an estimate may be conservative if it is known to be less than the true **parameter** value (actually **biased**) and is intentionally quoted as such to avoid the risk of it being an overestimate. $\Rightarrow$ **safety margin**

consistency check   an **edit check** on data to ensure that two (or more) data values could happen in conjunction. Systolic blood pressure measurements must always be at least as great as diastolic measurements so, for any given patient, if the systolic pressure is greater than the diastolic, then the two measures are consistent with each other. It may be that neither is correct—but they are, at least, consistent. $\Leftrightarrow$ **plausibility check**

consistent   reproducible without upward or downward trends over time. Also two items (often data points) that could both occur simultaneously. $\Rightarrow$ **consistency check**

CONSORT   a set of guidelines, adopted by many leading medical journals, describing the way in which clinical trials should be described. It stands for Consolidation of the Standards of Reporting Trials. $\Rightarrow$ **structured abstract**

constant   not changing between subjects or across time

consumer's risk   the **probability** of committing a **Type I error**. $\Leftrightarrow$ **producer's risk**. $\Rightarrow$ **regulator's risk**

contingency table   a cross-classification of subjects by two or more **categorical variables**. The simplest form is the **two-by-two table** (Table 3), in which each subject is cross-classified by two **binary variables**. The table has four **cells** (totals are not usually counted) and the number of items within each cell is called the **cell frequency**

continual reassessment method   a procedure for adjusting the dose given to successive subjects when the purpose of a study is to find the median dose that has some specified effect. $\Rightarrow$ **dose finding study**, **dose escalation study**

Table 3   Contingency table showing the distribution of gender by treatment group

	Treatment A	Treatment B
Male	58	63
Female	29	28
Total	87	91

continuity correction   an adjustment made in the calculations for some **significance tests** on **discrete data** to make a better approximation to the **test statistic** that is continuous. It generally involves adding or subtracting 0.5 to the difference between the observed and expected frequencies of data. In **two-by-two tables** it is often referred to as **Yates' correction**

continuous data   data that are not restricted to particular values (as in categories) but that can take an infinite number of values. Examples of variables that result in continuous data are age, height, weight, pulse. ⇔ **categorical data**, **discrete data**, **ordinal data**

continuous scale   the scale on which **continuous data** are measured

continuous variable   a characteristic of a subject that results in **continuous data**. For example age, height, weight. ⇔ **discrete variable**

contour plot   a graph that shows three dimensional data on a two dimensional surface. Two variables are depicted on the x **axis** and y **axis**; the third is depicted in the form of contours as would be seen on a map to show elevation (Figure 5)

contractor   a temporary employee, usually of professional rather than clerical status, taken on to perform duties that would otherwise be carried out by full time employees. Such people are often used to cover peaks in workload or periods of absence of permanent employees

contraindication   an **indication** for which a drug is specifically excluded

contrast   a more formal term for comparison. In its simplest form it is the difference in the mean value of a variable between two groups or the difference in the proportion of subjects with some particular characteristic in each of two groups. In more complex forms it may be a **weighted** difference between several groups. For example, in a study with three groups of subjects, two groups (A and B) treated with active treatments and a third group (C) treated with placebo, a simple contrast would be that between the two active products which is simply mean(A) − mean(B). We may wish to compare the active products together with the placebo group and so a more complex contrast would be $\{\frac{1}{2}$mean(A) + $\frac{1}{2}$mean(B)$\}$ − mean(C)

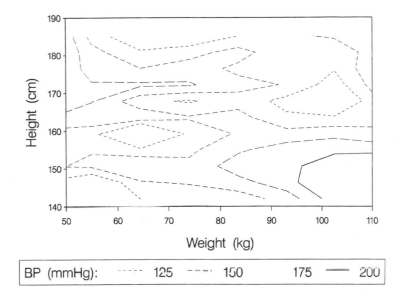

Figure 5 Contour plot. The heights and weights of 100 patients with ischaemic heart disease are used to try to predict systolic blood pressure. In general there is a tendency for higher blood pressure in the bottom right-hand corner: that is, the heavier people who are rather short (and therefore those that are most overweight) have the highest blood pressure

control a term used in **case-control studies** specifically intended to mean someone who does not have any disease. ⇔ **case**

control group the subjects assigned to receive the **comparator treatment**, or to receive no treatment

control treatment ≈ **comparator treatment**

controlled clinical trial more formal term for **clinical trial**. It clearly emphasises the 'controlled' aspect of a trial, although that should be inherent in the definition of clinical trial

controlled experiment a term similar to **controlled clinical trial**, except that it could refer to any kind of experiment, not just a clinical (or even medical) one. The aspect of control (and therefore inclusion of one or more **control groups**) is still emphasised

convenience sample   a sample of subjects. Whether or not a subject is selected for the sample is not based on any **random** process but merely on which people are conveniently available. ⇨ **haphazard sample**

coordinate   ≈ *xy* **coordinate**. Also means to ensure that several activities happen together (as they should do) or in sequence (if that is how they are intended to occur)

correlate   to assess how one variable changes as another changes. ⇨ **correlation**

correlated samples t test   ≈ **paired *t* test**

correlation   the degree to which two variables are associated with each other. **Positive correlation** (≈ Figure 34, **scatter plot**) implies that as one variable increases so does the other; **negative correlation** implies that as one variable increases the other decreases. Note that no **causality** is implied

correlation coefficient   the statistical measure of **correlation**, denoted *r*. ⇨ **coefficient of determination**

correlation matrix   a square **matrix** whose values are the **correlation coefficients** between all pairs of several variables. An example of the correlation between five laboratory parameters is shown in Table 4

correlation table   ≈ correlation matrix

cost benefit ratio   the relative weighting of the cost of a medication to the benefit of that medication. Benefit may be defined in arbitrary ways to suit the context. ⇨ **cost effectiveness ratio**, **cost utility ratio**

cost effective   generally meaning good value for money; the benefit outweighs the cost

cost effectiveness ratio   the relative weighting of the cost of a medication to the clinical effectiveness of that medication. ⇨ **cost benefit ratio**, **cost utility ratio**

cost function   an equation that calculates the total cost of treating a patient. It will typically include positive values (drug costs, pharmacy costs, hospital costs, productivity lost from work, etc.) but sometimes also negative costs (reduction in number of days spent in hospital, increased productivity from early return to employment, etc.)

cost minimisation   the approach of evaluating the optimum amount to spend in order to minimise the overall **cost function**

cost utility ratio   the relative weighting of the cost of a medication to the **utility** of that medication. Utility is the overall benefit as assessed by any and all diverse measurement scales including medical, financial, quality of life, etc. ⇨ **cost benefit ratio**, **cost effectiveness ratio**

COSTART   a dictionary of **adverse event** terms. COSTART stands for

Table 4   Correlation matrix of biochemistry parameters in 100 healthy subjects

	Urinary creatinine	Urinary calcium	Serum phosphate	Serum creatinine	Serum calcium
Urinary creatinine	1.0	0.41	−0.03	0.03	0.08
Urinary calcium	0.41	1.0	−0.06	0.00	0.11
Serum phosphate	−0.03	−0.06	1.0	0.07	0.15
Serum creatinine	0.03	0.00	0.07	1.0	−0.05
Serum calcium	0.08	0.11	0.15	−0.05	1.0

Coding Symbols for Thesaurus of Adverse Reaction Types. ⇨ **MedDRA, WHO-ART**

count   to determine how many (of something) exist or how many times a certain type of event has occurred

covariance   a statistical measure of how two variables vary together. ⇨ **correlation, variance**

covariate   a variable that is not of primary interest but which may affect response to treatment. Common examples are subjects' **demographic data** and baseline assessments of disease severity

Cox model   ≈ **Cox's proportional hazards model**

Cox's proportional hazards model   a statistical method for comparing **survival times** between two or more groups of subjects that also allows adjustment for **covariates**. The model assumes **proportional hazards**. ⇨ **Cox–Mantel test, accelerated failure time model**

Cox–Mantel test   a statistical method for comparing **survival times** between two groups. ⇨ **Cox's proportional hazards model**

cream   a mixture of **ointment** (such as paraffin, lanolin, etc.) and water used as a **vehicle** for delivering topical treatment. ⇔ **gel, lotion**

credible interval   a form of a **confidence interval** used in the context of **Bayes' theorem**. ⇨ **highest density region**

critical appraisal   the set of skills (and judgements) needed to evaluate evidence. ⇨ **evidence-based medicine**

critical data   the most important data that will be used to draw conclusions from a study relating to the most important **objectives**

critical region   the values of a **test statistic** (such as in the *t* test or **chi-squared test**) that lead to rejecting the **null hypothesis** at a given **significance level**

critical value   the value of a **test statistic** (such as in the *t* test or **chi-squared test**) that is the boundary between where the **null hypothesis** is rejected and not rejected at a given **significance level**

Cronbach's alpha   a measure of **internal consistency** in a psychological test

cross-classification   ≈ **contingency table**

crossed factors   the opposite of **nested factors**. When every category of one variable also contains every category of another. ⇨ **factorial design**

crossover design   a study where each subject receives (in a **random** sequence) each study medication. After receiving Treatment A, they are 'crossed over' to receive Treatment B (or vice versa). This is the simplest form of crossover design and is called the **two period crossover design**. ⇔ **parallel group design**

crossover study   a study that is designed as a **crossover design**

cross-product ratio   ≈ **odds ratio**

cross-sectional   considering a single moment (or separate moments) in time without regard for any trend across time. ⇔ **longitudinal**

cross-sectional analysis   the analysis either of a **cross-sectional study** or of data as if they were collected in a cross-sectional study. ⇔ **longitudinal analysis**

cross-sectional study   a study that examines data at one particular point in time (either in the sense of 'all ten-year-old children' or 'everybody on 1st January') and does not consider **within subjects effects**

crude estimate   any estimate of a parameter that is an **unadjusted** estimate

crude rate   an **unadjusted** rate. Generally, simply the observed number of subjects experiencing a specific event divided by the total number of subjects exposed and potentially at risk of that event

cumulative frequency   a running total. For example, if we count the number of deaths per day (the daily frequency), then the total number of deaths from the beginning of a study to any particular day is the cumulative frequency

cumulative frequency distribution   the **distribution** of **cumulative frequencies**

cumulative hazard rate   the accumulation of the **hazard functions** at all times from time zero up to a specified time point

cumulative meta-analysis   a meta-analysis that shows continuing updated estimates of treatment effect after each of the studies was completed. It does not simply show one overall result taking account of all studies, regardless of when they were carried out

curriculum vitae   a person's educational and employment history, usually including all other relevant experience and any publications to which they have contributed

curve   a smooth line or surface drawn though a set of data points. The term can strictly be used to describe a line (in two dimensions) or a surface (in more than two dimensions) that is straight as well as one that bends

curvilinear regression   a **regression model** that fits a curve to data. In this context, curve is generally taken to exclude a straight line. ⇨ **linear regression**

cutoff design   a method of treatment assignment based on a **baseline** measurement. All subjects with values below some cutoff point (defined as those with good prognosis) are assigned to the control group; subjects with values of the baseline measurement in the middle of the range are not included in the study; and all subjects with values of the baseline measurement above another cutoff point (those with poor prognosis) are assigned to the experimental group. ⇨ **regression discontinuity design**

cutoff point   a value on an **ordered scale** (possibly a **continuous scale**) where a change of decision is made. For example, patients with systolic blood pressure above 180 mmHg may be included in a study; those with values less than, or equal to, 180 mmHg are not included: 180 would be called the cutoff point

cutpoint   a point along a line or on a surface where the slope changes abruptly rather than smoothly. ⇨ **changepoint model, cutoff point**

cyclic   ≈ **cyclic variation**

cyclic variation   **systematic variation** over a course of time. A **circadian rhythm** is one type of cyclic variation

cytotoxic   a drug that is poisonous to certain types of cells. Frequently used in cancer treatment

data information of any sort, whether it be numerical, alphabetical, judgements, estimates or precise measurements. ⇨ **binary data, categorical data, continuous data, discrete data, ordinal data**

data analysis the process of summarising data, either to draw conclusions or simply to describe a process

data and safety monitoring committee a group of people who regularly review **accumulating data** in a study with the possibility of stopping the study or modifying its progress. A study may be stopped, or changes made to it, if clear evidence of efficacy is seen or if adverse safety is observed in one or more treatment groups

data audit an **audit** of the quality, source and integrity of data

data centre the place where data are gathered and the **data management** tasks completed. It is a term particularly relevant to **multicentre studies**. **Single centre studies** may have the data centre at the same place as the patients are seen or somewhere different

data cleaning the process of finding errors or possible errors in data, checking them and, if appropriate, correcting them. ⇨ **clean data, dirty data**

data coding assigning data into categories. For example, classifying **adverse events** into groups according to which part of the body is affected or classifying **concomitant medications** into **generic names** rather than **trade names**

data collection form ≈ **case record form**

data collection protocol specific, detailed instructions for how data are to be collected and recorded

data coordinating centre ≈ **data centre**

data dependent stopping making the decision to stop recruitment to a study (or possibly follow-up in a study) based on data already observed. ⇨ **interim analysis**

data dredging analysing data without regard to accepted scientific and statistical principles in order to find some aspect that will be of interest. Also referred to as 'fishing expeditions' because of the analogy of dipping a fishing rod into dark water and pulling out various items of rubbish, but rarely fish!

data driven analysis   making decisions on which analyses should be carried out based on the observed data. ⇨ **post hoc analysis**

data editing   ≈ **data cleaning**

data entry   typing data into a computer. This may be done directly by the subject, by the treating doctor or investigator or, more usually, copied from a **case record form** at a **data centre**. ⇨ **single data entry, double data entry**

data field   an individual item of data. The term is most often used in referring to data on a **case record form** or on a computer

data file   a highly structured and well organised collection of related data. The term could be used about paper files (including a **case record form** or a large number of case record forms) but is generally reserved for data on a computer

data item   ≈ **data field**

data management   the discipline of collecting and filing data in an ordered fashion to facilitate subsequent retrieval and analysis. Although the term can refer to the management of paper files, most activity in data management usually revolves around storage of electronic versions of data on a computer

data manager   the person with the responsibility for ensuring **data management** is properly carried out

data monitoring   the process of reviewing data being collected to ensure it is of high quality and complete. ⇨ **quality control**

data monitoring committee   ≈ **data and safety monitoring committee**

data monitoring report   a report on the quality and completeness of data

data processing   the steps involved in computerisation and particularly **data management** in a computer system

data query   a question raised about the validity or correctness of an item of data

data reduction   the process of summarising data, particularly using **summary measures** or coding **continuous data** into **categorical data**

data screening   the process of looking at and reviewing data to check their plausibility and completeness

database   an electronic version of a set of data, held on a computer

database management system   any piece of **software** for handling data. This includes **data entry** as well as production of tables, listings, etc.

dataset   ≈ **data file**

death rate   the number of people dying in a specified time interval divided by the number alive at the beginning of that time interval. ⇨ **case-fatality rate**

debug to find errors in computer programs and correct them

decile each of the tenth (10th, 20th, 30th, etc.) **centiles**

decimal a number recorded in whole units and in tenths, hundredths, thousandths, etc. For example, weight measured in kilograms and grams is expressed as a decimal number but weight measured in pounds and ounces is not. Sometimes the word is used just to refer to the part of the number less than 1 (only the numbers that come after the decimal point)

decision function a mathematical function that describes which decision to make, based on a given set of circumstances. ⇨ **decision rule, decision tree**

decision rule this term can be either a synonym for a **decision function** or a less technical, written, description of what decision to make based on a given set of circumstances. The rule may sometimes be depicted as a **decision tree**

decision theory the general theory of how to make optimal decisions

decision tree a diagram, resembling that of a family tree, to guide which decision (or sometimes which conclusions) should be drawn from a set of criteria (Figure 6). ⇨ **decision rule**

Declaration of Helsinki a set of ethical guidelines for the conduct of research on humans. It was first agreed in 1964 by the World Medical Association and has been revised subsequently in Tokyo (1975), Venice (1983), Hong Kong (1989) and South Africa (1996)

decrement to decrease in value. ⇔ **increment**

deduce to draw a conclusion of a specific result based on broader examples. ⇔ **induce**

deduction ≈ **deduce**

deductive inference the process of drawing conclusions based on **deduction**

deductive reasoning see **deductive inference** but note that 'reasoning' is a broader term than 'inference'

default an assumed state unless a positive reason can be given to accept an alternative state. For example, in **significance testing**, by default the **null hypothesis** will be accepted unless evidence exists to refute it

definitive study a study that is generally agreed to provide the answer to a question with no room for doubt. The term 'definitive' is usually used to describe a study that has already been completed. The term **confirmatory study** is more often used of a study that it is planned to undertake

degree of belief often used as an informal interpretation for a *P*-value. It is a measure (either on a **probability** scale or an informal, intuitive, scale) of the **strength of evidence** about a particular **hypothesis**

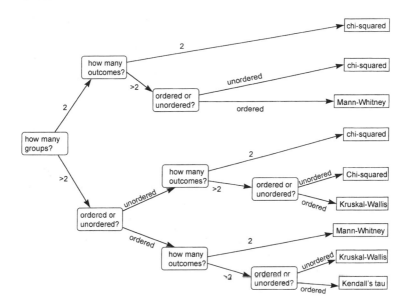

Figure 6   Decision tree. A way to make a choice of a simple statistical significance test for comparing groups of categorical data. Each of the boxes with rounded corners is called a 'node'; each of the arrows is called a 'branch'

degrees of freedom   a statistical term to describe the number of independent pieces of information that there are for a **statistic**. In **chi-squared tests** of **two-by-two tables**, there is one degree of freedom, the sample mean of n data points has $n-1$ degrees of freedom

delivery device   the medium used for getting active product into the body. **Tablets, ointments, injections**, etc. are all delivery devices. $\Rightarrow$ **vehicle**

delta (δ)   usually used as the symbol to describe the 'true' size of an effect. In particular it is used in planning studies to describe the smallest **clinically significant difference** to detect. It is more often used to describe a difference in means but can also be used to describe a difference in rates or proportions. The symbol d is often used to describe the observed value of δ

demographic data   data on subjects' age, height, weight, etc. The term can be used to describe any **baseline** characteristics of subjects including the baseline measurements of the **primary endpoint** variable but is more often reserved for measurements that are not aspects of the disease. There is no clear distinction between which data are disease related and which are not; clearly in a study of weight loss, subjects' weight would be both demographic data and important data describing the state of disease

demographic variable   any variable that is **demographic data**

demographics $\approx$ **demographic data**

demography   the study of **vital statistics** of populations

denominator   in a fraction, such as $\frac{1}{2}$ or $\frac{3}{4}$, the denominator is the number on the bottom line of the fraction (in these cases 2 and 4, respectively). $\Leftrightarrow$ **numerator**

density function   the mathematical function that gives the probability that a **random variable** is equal to any given value. $\Rightarrow$ **distribution function**

dependent samples *t* **test** $\approx$ **paired** *t* **test**

dependent variable   in any sort of statistical model, but most commonly in **regression models**, the dependent variable is the one we are trying to predict from the **independent variable**(s). In most cases, the dependent variable is the **efficacy variable**

derived variable   data values that are calculated or formed from other data. For example, subjects' age might be calculated (or derived) from the **visit date** minus the date of birth; age would then be called a derived variable

descending order   data sorted so that the largest value is written first, the smaller values later and the smallest value last. Most easily described in terms of **numeric** data but special rules can be applied to **alphanumeric** data. $\Leftrightarrow$ **ascending order**

descriptive statistics   summaries of data that do not try to draw conclusions but which just describe the data. Most often used for **continuous data**. Common descriptive statistics include the **mean**, **standard deviation**, **minimum** value, **maximum** value, **mode**, **median**, **quartiles** and **confidence intervals**. $\Leftrightarrow$ **inferential statistics**

descriptive study   one that aims to describe a phenomenon or a group of individuals. The analysis of data from such studies generally uses **descriptive statistics** rather than **significance testing** or **inferential statistics**

design   the plan for a study with particular reference to whether it is a **parallel group design** or **crossover design**. The term should, however, be thought of very broadly to encompass the number of subjects to be included, the number of visits, the number of investigators taking part, **strata**, **blocking**, methods of **randomisation**, etc.

design effect   the effect caused by a **design variable** in a study. Such effects would, hopefully, be advantageous but they may be negative or neutral. In a study where the **randomisation** was **stratified** by gender, an observed difference in treatment effect between males and females would be called a design effect (because stratification was part of the design of the study)

design variable   any variable that contributes to the **design** of a study, often because of **stratification** according to values of the variable

deterministic   a process that is guaranteed to give the same result repeatedly, with no unexplainable (random or otherwise) variation

deviance   a statistical measure of how much a set of data differs from a perfect fit to a **model**. In the simplest case of a model with **normally distributed residuals**, the deviance is equal to the **residual sum of squares**. ⇨ **variance**

deviate   a variable that takes the values of the difference between another variable and a chosen **reference value**, such as the mean

deviation   a measure of how far values of a variable lie from a chosen reference point. ⇨ **average absolute deviation, standard deviation**

device   see **delivery device, medical device**

diagnosis   the decision that is reached regarding the disease a patient has

diagnostic test   a test (physical, mental or, more commonly, biological) that is used to definitively assess whether or not a subject has a particular disease. ⇔ **screening test**

diagram   a line drawing, usually to show the relative positions (physically or in time) of a set of objects or activities

diary card   usually a paper system for subjects in a study to record **symptoms, adverse events** or other data on a daily basis, often at home and generally not under the direct supervision of any medical personnel

dichotomous data   ≈ **binary data**

dichotomous outcome   ≈ **binary outcome**

dichotomous variable   ≈ **binary variable**

difference   the value obtained by subtracting one value from another. This may be on an individual subject basis, for example calculating the difference between a subject's pulse at baseline and the same subject's pulse after treatment (≈ **change from baseline**), or it may be on a group basis, for example calculating the difference between the mean of all subjects' heart rates in one treatment group and the mean of all subjects' heart rates in a control group

difference study   a term used rarely, except to differentiate from an **equivalence study** or a **noninferiority study**. A study where the **null hypothesis** is that there is no difference between treatments and the

alternative hypothesis states that there is a difference. The intention of the study (or **objective**) is usually to show that two (or more) treatments have different effects. ⇨ **superiority study**

diffuse prior   ≈ **vague prior**

digit   any numeral between zero and nine. For example, the number 57 contains two digits: 5 and 7

digit bias   ≈ **digit preference**

digit preference   when recording numerical data, there is often a preference (intentional or unintentional) to round the last digit. For example, birth weight measured in grams will often be recorded to the nearest 10 grams; there is said to be a preference for zeros. Blood pressure measured in millimetres of mercury will often be recorded to the nearest 5 mm or the nearest 2 mm; values such as 73 mmHg (where the last digit is not a multiple of 2 or of 5) tend to be recorded less often than would be expected by chance

dimension   one of any number of variables that describe a subject. The term is most often used in connection with plotting data. When two variables are measured and plotted, there are two dimensions. When there are three variables plotted, three-dimensional graphs can be plotted (with some difficulty). More than three dimensions can be thought about but cannot easily be plotted. ⇨ **multivariate data**

direct access   in computing terms this refers to the method of accessing data on a physical storage device such as a **disk**. (⇔ **sequential access**.) The term also applies to **source data verification**, where the person reviewing **source data** is allowed to see the source data for themselves, rather than indirect access where the values of source data have to be requested through a third party

direct contact   the contact of one person with another that potentially passes on an **infectious** disease. ⇔ **indirect contact**

direct cost   actual (financial) costs that are incurred in treating patients. These include the cost of drugs, the cost of occupying a hospital bed, etc. ⇔ **indirect cost**. ⇨ **pharmacoeconomics**

direct effect   ≈ **main effect**

direct relationship   the case when the relationship between two variables is **linear**, so that plotting one variable against the other variable shows a rough fit to a straight line. The term is often further restricted to the case when the **correlation** is positive—such as in Figure 34 (≈ **scatter plot**)

directional hypothesis   a **hypothesis** which specifies that one treatment is equal to, or better than, another treatment. In general, the **alternative hypothesis** is stated that one group is different from another, which

Table 5   Cross-classification of paired data to show the discordant pairs. The response to each treatment (in this example) is graded simply as 'good' or 'bad'

Treatment B	Treatment A	
	Good	Bad
Good	55	47
Bad	13	24

could allow it to be either better or worse. ⇨ **one sided hypothesis**

dirty data   data that contain errors, or data that may contain errors and have not yet been fully reviewed and validated to find those possible errors. ⇔ **clean data**

discordant pair   in a study where subjects are assessed on two different occasions or by two different measuring devices and the variable measured is **binary** (for example, disease present or absent), the data may be summarised in a **two-by-two table**. The discordant pairs are those pairs of observations where the two measurements do not agree with each other (Table 5). In this example, 47 patients and 13 patients represent the discordant pairs. ⇔ **concordant pair**

discrete data   data that may take only a fixed set of values. This includes **categorical data** but also extends to data in the form of **counts**, for example where only whole numbers of items can be counted. ⇔ **continuous data**

discrete variable   a variable that can result only in **discrete data** values. ⇔ **continuous variable**

discussion   that part of a **final report** that addresses the validity of the results by considering the appropriateness of the study design, the success (or otherwise) of its implementation, quality of the data, consistency of results across different **outcome variables** and in the light of other studies. ⇨ **conclusion**

disease profile   the set of **signs** and **symptoms** (and their severity) that either characterise a disease (and therefore may help with **diagnosis**) or describe the severity of disease for an individual patient

disk   a device for storing data on a computer or in a computer readable form. Traditionally these have been magnetic devices but optical devices (compact discs, etc.) are becoming very common

diskette   virtually synonymous with **disk**. Some people use the term diskette to refer to 'small' floppy disks that can be carried around (usually for use with personal computers) rather than larger hard disks that are kept permanently inside the computer

dispersion   a term used almost synonymously with **variability** (as in variation of data). ⇨ **variance**

distributed data entry   a system of entering data onto a variety of computers, possibly spread around the world, to form a **distributed database**. ⇨ **remote data entry**

distributed database   rather than all the data relating to a study being held on a single computer, a distributed database allows different parts of the data to be held on different computers. The different computers are all linked together by a **network** so that it is not obvious to the user that the database is distributed

distribution   a general term covering either **frequency distribution** or **probability distribution**, depending on the context

distribution free method   ≈ **nonparametric method**

distribution function   the mathematical function that gives the probability that a **random variable** is less than any given value. ⇨ **density function**

divisor   ≈ **denominator**

doctors and dentists exemption (DDX)   an exemption similar to a **clinical trial exemption certificate (CTX)** but one that is issued to a doctor or dentist, not to a pharmaceutical company

documentation   written evidence to confirm the activities that have been undertaken in a study and the standards to which a study has been managed

dosage regimen   the dose, timing and method of giving medication to a patient. ⇨ **treatment regimen**

dose   the amount of drug that is given

dose effect relationship   ≈ **dose response relationship**

dose escalation study   a study in which successively higher doses of a drug are given to subjects. This may be done either by administering a dose to an individual and, if there are no **adverse events**, by increasing the dose for that individual until adverse events are seen or by giving a dose to a small number of subjects and, if no adverse events are seen, giving a subsequent group of subjects a higher dose, and so on, until adverse events are seen. ⇔ **dose ranging study**

dose finding study   a study to find the best dose ('best' according to an agreed definition) of a drug

dose ranging study   a study of different doses of a drug but, in contrast to a **dose escalation study**, the doses being compared are not investigated in an escalating manner

dose response   ≈ **dose response relationship**

dose response relationship   how the effect of a drug changes with dose

dose titration study   ≈ **dose escalation study**

dosing schedule   ≈ **dosage regimen**

dot chart   ≈ **scatter plot**

dot plot   ≈ **scatter plot**

double blind   a study where the subjects and the investigators are **blind** to the **treatment allocation**. ⇔ **single blind, triple blind**

double data entry   a strategy where data from **case record forms** is entered (typed) into a computer twice and the two typed files compared. This helps to reduce the number of typographical errors and errors of interpretation of poor handwriting. ⇔ **single data entry**

double dummy   a method of blinding where both treatment groups may receive **placebo**. For example, one group may receive Treatment A and the placebo of Treatment B; the other group would receive Treatment B and the placebo of Treatment A

double entry   ≈ **double data entry**

double mask   ≈ **double blind**

doubly censored data   data that are both **left censored** and **right censored**. Right censored data is quite common; left censored data is less common; doubly censored data is rare

doubly censored observation   ≈ **doubly censored data**

download   copying files (data or programs) from a central computer to a local computer. ⇔ **upload**

dropin   the opposite of **dropout**. Dropins to clinical trials are not common but, when they occur, may result in **left censored observations**

dropout   the case where a subject stops participating in a study before they are due to according to the study protocol. A more polite term is **early withdrawal**

drug   a pharmaceutical preparation. The term is often used very broadly and loosely to include **placebo**. ⇨ **biologic, phytomedicine**. ⇔ **product**

drug accountability   the process of checking what has happened to all **study medication**. This includes checking stocks in a **pharmacy**, counting individual subjects' tablets, weighing tubes of ointment, etc.

drug company   ≈ **pharmaceutical company**

drug industry   ≈ **pharmaceutical industry**

drug interaction   the effect sometimes produced when more than one product is used simultaneously. The effect is either more than or less than the sum of the individual effects. The term is most commonly used in connection with **adverse reactions**, caused by different products combining in the body, rather than with extra **beneficial effects**

drug metabolism $\approx$ **metabolism**

drug reaction   any response to a product, either beneficial or unwanted, but usually reserved for unwanted effects. $\Rightarrow$ **adverse event, adverse reaction**

drug trial $\approx$ **clinical trial**

dry run   similar to a **pilot study**. Trying a process under artificial conditions to determine if it will work properly in a real setting

dummy loading   a method of **blinding** treatments when they involve different **dosage regimens**. $\Rightarrow$ **double dummy**

dummy report $\approx$ **ghost report**

dummy table $\approx$ **ghost table**

dummy variable $\approx$ **indicator variable**

Duncan's multiple range test   a **multiple comparison test** for comparing the mean value of a variable between more than two groups

duration of action   the length of time that a treatment gives any benefit

duty of care   the requirement that doctors must care for their patients and that this duty must take priority over such things as research projects

dynamic allocation   a **randomisation** method that changes the probability of assignment from one group to another as the study progresses. The probabilities are changed either as a consequence of **efficacy** and **adverse event** data emerging or to maintain balance for **prognostic factors** across the groups. $\Rightarrow$ **minimisation**

early stopping the practice of stopping **recruitment** into a study before reaching the maximum target **sample size**. This may be in a **sequential study**, after a formal **interim analysis** or for purely practical reasons that are independent of efficacy or safety results

early stopping rule a statistical rule that allows a study to stop recruitment after an **interim analysis**. Unless such rules are used the *P*-value associated with testing the **null hypothesis** is generally biased (it is too small). Early stopping rules allow for this and help to calculate the correct *P*-value

early withdrawal when a subject leaves a study earlier than is routinely allowed for in the **protocol**. Typical reasons include the onset of unacceptable **adverse events** and voluntary withdrawal. In studies where death is not the **endpoint**, a death might also be included as an early withdrawal

edit the process of changing data or text in a datafile or in a text document (usually one held on a computer)

edit check a term that covers all types of checks that may be put on data, including **consistency checks, plausibility checks, range checks**

edit query a question raised by an **edit check**. The relevant data would then be checked and appropriate corrective action taken if necessary

effect this term is often misinterpreted as being the **change from baseline** in some measurement (blood pressure, for example) during the period of an intervention. Strictly speaking, 'effect' should always be a relative measure, such as the extra change in blood pressure over that produced by the comparator treatment. If the mean blood pressure in a treatment group falls by 15 mmHg and in a comparator group it falls by 5 mmHg, then the effect is the difference between these two values—10 mmHg. Similarly, the effect of gender is defined as the difference in mean response between males and females; the effect of study centre is defined as the difference in mean response between participating study centres. When the outcome of interest is not a mean, the term 'effect' has the same limitation in that it should always be the difference between two groups; this can be measured as the difference between two **proportions,** or the **odds ratio,** or the difference in median **survival times,** etc.

effect modifier   ≈ covariate

effect size   strictly this should simply be the size of an **effect** but conventionally it is taken to be the size of the effect divided by the **standard deviation** of the measurements. An effect size of one indicates a difference between two means equal to one standard deviation; this is generally considered to be quite a large effect. Effect sizes of about 0.5 are moderate and effect sizes 0.1 or lower are considered very small

effective sample size   the more variation there is in data then, generally, the larger the **sample size** required to show a treatment effect. However, if, for a given sample size, there is more (or less) variability in the data than anticipated it is as if there is a reduced (or increased) sample size. The sample size that would have been required, had the variability in the data been correctly assessed, is called the effective sample size. Variability in data can be increased due to **missing data** (⇨ **early withdrawal**) and errors in the data; it can be reduced by modelling the data using extra **covariates.** ⇨ **relative efficiency**

effectiveness   the extent to which a product works in the patients to whom it has been offered. This is slightly different from 'efficacy', which can be measured in those who were actually treated. 'Efficacy' relates to **explanatory studies,** 'effectiveness' to **pragmatic studies**

efficacy   the desirable effect of an **intervention.** ⇔ **safety.** ⇨ **adverse event**

efficacy data   any data relating to the **efficacy** of a treatment. ⇔ **safety data**

efficacy population   ≈ **per protocol population**

efficacy review   an **overview** or **meta-analysis** of efficacy data. ⇔ **safety review**

efficacy sample   ≈ **per protocol population**

efficacy study   a study intended primarily to demonstrate **efficacy** rather than **safety.** Often the same as a **Phase III study**

efficacy variable   a variable that is a measure of **efficacy.** ⇔ **safety variable**

efficient   a process that makes good use of resources and is not wasteful. This is also a statistical term referring to methods of estimating **parameters**: in general it is desirable to have **efficient estimators** because these may require smaller **sample sizes**

efficient estimator   an estimate of a **parameter** that is **efficient**

eighty–twenty rule   an informal rule which suggests that most benefit (about 80%) can be achieved with minimal effort (about 20%); and, conversely, that the last 20% of benefit needs 80% of the effort

elective treatment   a treatment that a patient chooses to have rather than one that is assigned by **randomisation** or one that is mandatory on medical (or other) grounds

electronic database $\approx$ **database**

eligibility criteria $\approx$ **inclusion criteria**

eligible   a subject that meets all the eligibility criteria ($\approx$ **inclusion criteria**)

elimination   the process by which a drug is excreted from the body or removed from the required site of action within the body. $\Leftrightarrow$ **absorption, clearance**

elimination rate constant   once a drug has been completely absorbed into the body, this is the rate of **elimination** (which, for many drugs, is approximately constant)

empirical   observed (particularly in relation to curves, **distributions**, etc.) $\Leftrightarrow$ **fitted value**

empirical Bayes   **Bayesian** methods that require the **prior distribution** to be based on data. $\Leftrightarrow$ **subjective Bayes**

empirical distribution   the observed **frequency distribution** of data. $\Leftrightarrow$ **probability distribution**

empirical result   a result based on data (or facts) rather than one based on theory

empty cell   in a **contingency table**, a **cell** that contains no observations

end of study $\approx$ **end of treatment**

end of treatment   the time at which subjects are either supposed to stop taking treatment (according to a **protocol**) or actually do stop taking treatment (if, for example, they were an **early withdrawal**)

end of treatment value   the value of a variable at the **end of treatment visit**

end of treatment visit   the **visit** at which subjects are supposed to stop taking treatment (according to the **protocol**), actually do stop taking treatment or withdraw from a study

endemic   a disease that is always present in a certain proportion of the population in a given geographical area. The term is usually used when considering the frequency of extra cases of the disease

endpoint   a variable that is one of the primary interests in a study. The variable may relate to **efficacy** or **safety**. The term is used almost synonymously with **efficacy variable** or **safety variable** but not, for example, with **demographic variable**

enrol   to recruit a subject, or subjects, into a study

enrolment   the number of subjects that have been enrolled into a study

enrolment period   the time (often measured in months or years) during which subjects are enrolled into a study

enteric coating   a coating (often made of gelatine) used on a **tablet** or **capsule** to prevent it being destroyed by acid in the stomach

entry criteria   ≈ **inclusion criteria**

epidemiological study   a study using the methods of **epidemiology**. This includes clinical trials but also **case-control studies**, **cohort studies**, **natural experiments**, surveys, etc.

epidemiologist   one who studies or practices **epidemiology**

epidemiology   the study of health and disease in **populations**, including aetiology, natural course and treatments. Clinical trials are considered by many to be one of the methods of epidemiology

episode   the occurrence of an event. In some studies the **primary endpoint** or primary **efficacy variable** may be the number of times an event happens (the number of episodes of that event)

equal allocation   allocating the same number of subjects to each treatment. ⇔ **unequal allocation**

equal randomisation   ≈ **equal allocation**

equation   a set of mathematical symbols and instructions for performing calculations

equipoise   the state of having an indifferent opinion about the relative merits of two (or more) alternative treatments. Ethically, a subject should only be randomised into a study if the treating physician has no clear evidence that one treatment is superior to another. If such evidence does exist then it is considered unethical to randomly choose a treatment. If the physician is in a state of equipoise, then randomisation is considered ethical

equipotent   having equal **potency** and therefore having equal effects (positive or negative). ⇨ **equivalent**

equivalence   the situation where two treatments show equal **effects**

equivalence study   a study whose primary aim is to demonstrate that two treatments are **equivalent** with regard to certain specified **parameters**. Most studies are designed to show that one treatment is better than another; these are sometimes referred to as **difference studies** to emphasise the contrast with equivalence studies. ⇨ **noninferiority study**

equivalent   having equal **effects** (positive or negative). ⇨ **equipotent**

erect   standing. ⇨ **prone**, **supine**

error   a mistake. Sometimes the term is used to describe the discrepancy between an observed data value and the true value. In these situations, the term is used with reference to the **variance**, as in, for example, **error term**, **error variance**

error band   an informal term to describe an interval around an **estimate**

that **semiquantitatively** describes the uncertainty of the estimate of the parameter. ⇨ **interval estimate**

error bar an informal term, similar to **error band** but where the interval is shown on a graph. There is no fixed convention for the length of these 'bars' but they are typically one **standard error**, one **standard deviation**, two standard errors or two standard deviations. If error bars are used, their precise definition should be given. ⇨ **box and whisker plot**

error mean square   ≈ **residual variance**

error of the 1st kind   ≈ **Type I error**

error of the 2nd kind   ≈ **Type II error**

error of the 3rd kind   ≈ **Type III error**

error sum of squares   ≈ **residual sum of squares**

error term   ≈ **residual variance**

error variance   ≈ **residual variance**

errors in variables model   in many situations it is assumed that, although a **response variable** may be measured with uncertainty (because it has some **residual variance**), the predictor variables, or **covariates**, do not have any uncertainty in their measurement. This may often not be the case, and if it is not the relationship between the covariates and the response will be biased: **positive relationships** will be estimated as larger than they should be whilst **negative relationships** will be estimated to be smaller than they should be. If the variances of the covariates can be estimated, then an adjustment can be made to the estimated relationship with the response variable. A model that makes this adjustment is called an errors in variables model

essential documents   a regulatory term describing the documentation that is required to support the data from clinical trials. It includes the protocol, case record form, names and affiliations of all staff involved, including their curricula vitae, the source and quality assurance statements of the products involved, etc.

essential requirements   ≈ **essential documents**

estimable   a **parameter** that can be **estimated** from a given experimental design. Some complex **crossover studies** and **factorial studies** may intentionally include some parameters of lesser importance that cannot be estimated, in order to more efficiently estimate those parameters that are of greater interest

estimate   the value of a **parameter** that is calculated using data. It should always be remembered that exact answers to questions are rarely attainable because of **measurement error** and **random variation** in the variable we are trying to measure. The 'truth' is rarely known, the best we can usually do is to get estimates of it

estimated sample size the **estimate** of how many subjects must be enrolled into a study in order to meet the objectives of the study. ⇨ **sample size**

estimation the process of obtaining **estimates** of **parameters** from data

estimator a formula used to estimate a **parameter**

ethical a process or study that conforms to accepted guidelines and rules on **ethics**. ⇨ **Declaration of Helsinki**

ethical pharmaceutical a medicinal product that is available only with a doctor's prescription. ⇔ **over-the-counter drug**

ethics the discipline of describing behaviour, practices, thinking and moral values generally agreed to be acceptable to society. ⇨ **Declaration of Helsinki**

ethics committee   ≈ **research ethics committee**

ethnic origin a **demographic variable** encompassing place of birth, race, religion, and sometimes also native language. Often it is simply used to describe country or region of birth of a subject

evaluable subject one who conforms to the study **protocol** sufficiently well to be included in the **per protocol population**. Often this means a subject who meets all the **inclusion criteria** for a study and none of the **exclusion criteria**. Sometimes the requirements may be made less stringent and only certain major inclusion criteria need to be fulfilled. Sometimes the requirements may be more stringent and a certain minimum time in the study may be required. The precise definition of evaluable is likely to be study dependent and should be described in the protocol and study report

event a **binary variable** that is an **outcome** that may or may not occur for each subject in a study. Some events, if they do occur, can occur more than once. Events are more often considered as negative (≈ **adverse event**) but they may be positive aspects of a treatment

event rate the proportion of subjects who experience a particular **event** in a given **time interval**. Note that if the event can occur more than once for any given subject, as in **adverse events**, the event rate is still the proportion of subjects who experience that event; it is not a function of the number of events that occur

evidence based medicine a recent approach to patient management that relies on using the most rigorous data available to guide decisions on what treatments should be used and how they should be used. The forms of evidence preferred are usually (although not always) from **randomised** and **blinded** clinical trials and **meta-analyses**

exact statistical method   a statistical method for **estimation** and **significance testing** that does not make assumptions about the **distribution** of variables. Some exact methods are commonly referred to as **nonparametric methods** but the variety of exact methods currently being developed goes beyond what have traditionally been thought of as the nonparametric methods. ⇨ **parametric methods**

exact test   a statistical **significance test** using an **exact statistical method**

examination   a series of observations, usually undertaken to determine a **diagnosis** or to measure the progress of disease

exchangeability   a term used in the context of **bioequivalence** to encompass equivalence of all aspects of two products

excipient   the constituents of a product that are not active but help with the **formulation**. ⇨ **vehicle**

exclusion criteria   reasons why a subject should not be enrolled into a study. These are usually reasons of safety and should not simply be the opposites of **inclusion criteria**

excrete   to eliminate from the body, usually taken to mean via urine and faeces, but can also include sweat

excretion study   a study of the quantity, route, timing, etc. of drug being **excreted** from the body

executive committee   a small group of individuals representing a larger group, with the authority to make decisions regarding the design or conduct of a study. **Data monitoring committees** and **research ethics committees** could have a smaller group that meets more frequently than the main committee to pass through 'simple' decisions quickly or who meet on an infrequent basis to make 'major' decisions that have been discussed at length at a fuller committee

expectation $\approx$ **expected value**

expected frequency   the number of events that would be expected to occur within a set of constraints (usually the constraint is the **null hypothesis**). The term refers particularly to expected **cell frequencies** (as opposed to **observed frequencies**) in **contingency tables**

expected number $\approx$ **expected frequency**

expected outcome   in a statistical sense $\approx$ **expected value**. Otherwise the term is used in a general sense to refer to what **outcome** (or course of a disease) would generally be expected to occur. ⇨ **prognosis**

expected value   the value of a **parameter** that an **estimator** predicts. For example, the expected value of the **sample mean** is the **population mean**, although the expected value of the **sample variance** is not quite the **population variance**, there is a small **bias** (which can be corrected)

expedited report   a report that must be made very quickly. It usually refers to reporting **serious adverse events** to **regulatory authorities**, sometimes within two or three days of the event occurring

experiment   a general term that encompasses **preclinical studies, clinical trials, animal studies**, etc. It covers almost any form of practical research that involves **intervention**. ⇔ **observational study**

experimental design   all aspects of the design of an **experiment**. Sometimes the term is restricted to certain specialised statistical aspects of the design such as **blocking, replication** and **stratification**

experimental drug   ≈ **experimental treatment**

experimental error   ≈ **residual variance**

experimental treatment   usually the product that is of primary interest and that is being compared with the **comparator treatment**

experimental unit   this usually means each subject but is best thought of as the smallest unit that could be **randomised**. Even in studies that do not involve randomisation, it is still helpful to think in these terms. In **community intervention studies** the experimental unit might be an entire town; in other situations it could be a hospital ward or a General Practitioner's surgery. ⇨ **unit of analysis**

experimenter effect   ≈ **Hawthorne effect**

experimentwise error rate   the probability of making a **Type I error** when considering the overall result of a study. Note that, if a study has several **endpoints** to be analysed, even if one or more of those analyses may result in a Type I error the overall conclusion from the study could still be correct. ⇔ **comparisonwise error rate**. ⇨ **multiple comparisons**

expert report   a regulatory document that summarises the complete set of documents on the **safety** and **efficacy** of a **product** submitted for regulatory approval in a new indication

expert review   a review of documents, study results, etc. by an expert. ⇨ **peer review**

expert system   a computerised method of making decisions that is more complex than a simple **algorithm**; it is a method that is capable of 'learning' by building upon past decisions and their outcomes

expiry date   the date after which a product should not be used because its quality cannot be assured. ⇨ **shelf life**

explained variance   in a set of data there will usually be **variation** between data points. Some of this variation will be due to differences between subjects, differences between points in time, differences between treatments, etc. The variation that is due to such known causes is the explained variance; the variation that is due to unknown causes is called the

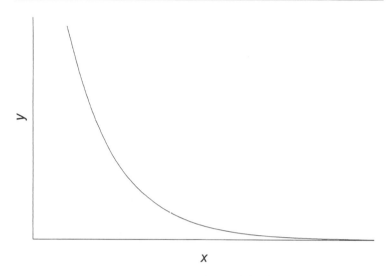

Figure 7 Exponential decay. For each equal sized change in the value of *x*, the value of *y* falls by the same proportion

residual variance, or simply the **variance**

explanatory study   a study that aims to find out if an intervention can work, given ideal circumstances, or to find the circumstances under which an intervention works. The analysis of such studies is usually by the **per protocol** approach. ⇔ **pragmatic study**

explanatory variable   ≈ **covariate**

exploratory data analysis   methods of reviewing data to find potential errors and to gain simple impressions of patterns that may exist or effects that may be happening. The methods are usually graphical and include **box and whisker plots, histograms, stem and leaf plots**

exploratory study   a study that aims to generate **hypotheses** rather than to definitively test them

exponent   in a mathematical equation of the form $y = x^z$ ('*x* raised to the power *z*') the parameter *z* is called the exponent

exponential   ≈ **exponential growth**

exponential decay   a quantity that is diminishing at an ever-decreasing rate (Figure 7). ⇔ **exponential growth**

exponential distribution   the **probability distribution** that describes the

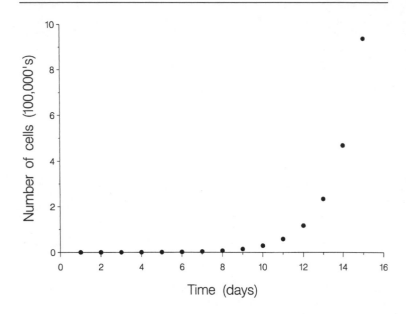

Figure 8 Exponential growth. Rate of growth of cancerous cells. The number of cells multiplies by the same factor after each additional day

time interval between randomly occurring events. It is an important distribution in the analysis of **survival data**

exponential growth   growing at an ever-increasing rate; for example, the number of cancerous cells in a tumour may double every week, or may increase tenfold every week (Figure 8). ⇔ **exponential decay**

exposed group   in a clinical trial this term is sometimes used to refer to the group receiving the **experimental treatment**. The term more naturally comes from **case-control studies** and refers to the **cases**

exposure   the extent (amount and length of time) for which a subject has received medication or other intervention (including possibly harmful interventions)

exposure variable   the variable that measures **exposure**

external consistency   a study whose results are applicable to, and match what is seen in, other studies and in **clinical practice**. All studies should obviously have this feature but many do not because they use **inclusion**

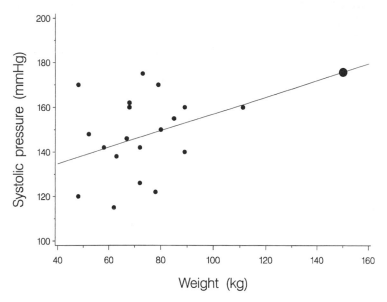

Figure 9 Extrapolation. A simple regression model predicting patients' systolic blood pressure from their weight has been used to predict what the blood pressure of a 150 kg person (the large dot) would be

criteria and **exclusion criteria** that are either different to those of other studies or are not reflected in clinical practice. ⇔ **internal consistency**. ⇨ **explanatory study**, **pragmatic study**

external validity ≈ **external consistency**

extrapolate either formally estimating, via a statistical **model**, or informally judging what results will occur outside the range of data actually collected and analysed. This may involve extrapolating to a wider **patient population** than has been studied, extrapolating from **animal studies** to judge what will occur in humans, etc. (Figure 9). ⇔ **interpolate**

extreme value the largest or smallest value in a set of data. Sometimes the extreme values (plural) are taken as several of the largest or smallest values

F

F **distribution**   a **probability distribution** used extensively for **significance testing** in **analysis of variance**. It is used to test whether two **variances** are equal but this can be put to use to compare the **means** across several groups

F **ratio**   ≈ *F* **statistic**

F **statistic**   the value of the **test statistic** calculated from an *F* **test**

F **test**   a statistical **significance test** based on the *F* **distribution**

F **to enter**   the value of an *F* **test** required as a **decision rule** to enter a variable into a **regression model** when using **forward selection** or **stepwise regression** methods. ⇔ *F* **to remove**

F **to remove**   the value of an *F* **test** required as a **decision rule** to remove a variable from a **regression model** when using **backward elimination** or **stepwise regression** methods. ⇔ *F* **to enter**

fabricated data   data that are not real and have been presented fraudulently. ⇨ **fraud**

face validity   a term usually used with reference to questions on a questionnaire. Face validity refers to whether a question seems to make sense to an expert in the field. It stems from the expression 'on the face of it'. ⇨ **external validity**, **internal consistency**

factor   another name for a **categorical variable**, usually (but not exclusively) one that is a **covariate** or a **stratification variable**, rather than one that is an **outcome variable**

factorial design   a study that compares two (or more) different sets of interventions. The simplest design uses Drug A versus Placebo A and Drug B versus Placebo B. Subjects will be randomised to one of four groups: Placebo A + Placebo B, Drug A + Placebo B, Placebo A + Drug B or Drug A + Drug B. This is a very efficient type of study because it not only allows the assessments of Drug A and Drug B in one study instead of two but also allows us to investigate the question of whether drugs A and B show any **interaction**

factorial study   a study of two or more interventions carried out in a **factorial design**

failure   the term sometimes used in place of **event** in **survival data**. It comes from studies of the time it takes for machine parts to cease working (or 'failing') but the term has been carried over to medical examples where we are looking at the time until an event such as death or relapse

failure time   the time until an event occurs, where the term **failure** has been used instead of **event**

false negative   the case when a test of some sort does not detect what it is supposed to detect. This can be a **diagnostic test** that fails to identify a patient who has a particular disease. The term is also sometimes used in **significance testing** to describe a **Type II error**. ⇔ **false positive**

false positive   when a test incorrectly detects something that is not real. In a **diagnostic test** this is identifying a patient as having a particular disease when they do not. In **significance testing** it is the same as a **Type I error**. ⇔ **false negative**

falsificationism   the act of falsifying data or results. ⇨ **fabricated data, fraud**

familywise error rate   ≈ **experimentwise error rate**

fatal   that which causes death. ⇔ **lethal**

feasibility study   ≈ **pilot study**

Fibonacci dose escalation scheme   a commonly used method of determining what doses of a drug should be used in a **dose escalation study**. The successively increasing doses follow a **Fibonacci series**

Fibonacci numbers   numbers that follow a **Fibonacci series**

Fibonacci series   a series of numbers that increase by successively adding the previous two numbers to get the next one. For example, $1, 1, 2 (= 1 + 1), 3 (= 2 + 1), 5 (= 3 + 2), 8 (= 5 + 3), 13 (= 8 + 5), 21 (= 13 + 8), 34 (= 21 + 13), \ldots$

fiducial inference   a method of statistical **inference** similar to **significance testing**. ⇨ **Bayesian inference, frequentist inference**

field study   a term used to describe a study that is not conducted in a hospital or similar type of well controlled environment but rather one that is carried out in general practice with patients free to carry on their normal daily activities. The analogy of an agricultural study being carried out either in a greenhouse-type environment or in a field where the climate and other environmental factors cannot be controlled is a good one. ⇨ **experiment**

figure   the term is used to refer to a number, or to a graph or diagram in a study report. The use can be confusing as different people assume it means different things

file   a physical or electronic (on a computer) place where documents and data are stored

final data analysis   the final analyses of a study that are reported. These may be done after various forms of **exploratory data analysis** have been completed

final report   another term for study report but the use of this term can be useful to distinguish it from an **interim report** or a draft of a study report

fine data   data measured with great accuracy. ⇔ **coarse data**

finite   having real bounds. The term is sometimes overused because most of what we do is finite. The use of the word can only really be justified if it genuinely contrasts with the possibility of being **infinite**

finite population   for the purposes of most statistical analyses, it is assumed that there are an infinite number of subjects to which the study results apply. This assumption is partly justified on the grounds that the possible set of subjects having the target disease includes all those with the disease today and all those who will have the disease in the future. In some situations this is not a sensible assumption and it must recognised that there is a finite number of subjects in the population to which our results can apply. ⇔ **infinite population**

first in man study   the first **Phase I study** undertaken with a new drug

first order interaction   ≈ **two factor interaction**

first pass metabolism   the **absorption** of drugs into the body when they pass through the liver

Fisher's exact test   a statistical **significance test** that is used for comparing proportions in **contingency tables**. It is used in preference to the **chi-squared test** when the sample size is small (often less than 30)

fishing expedition   ≈ **data dredging**

fit   to **estimate** the **parameters** of a **model** from data

fitted value   the **estimated** value of a **parameter** based on a **model**. ⇔ **observed value**. ⇨ **empirical result**

fixed combination therapy   a mixture of two (or more) drugs in one **formulation**. ⇔ **free combination therapy**

fixed cost   in **pharmacoeconomics** this refers to a cost that will remain the same however many patients there may be or in whatever way they may be treated. One might argue that the pharmacy department in a hospital needs to be open 24 hours a day whether it stores drugs for a certain disease or not: there is, therefore, a certain minimum fixed cost for this facility. ⇔ **marginal cost**, **per unit cost**, **variable cost**

fixed disk   ≈ **hard disk**

fixed effect   a **categorical variable** where the different **levels** of the **factor** are exactly the ones that we wish to draw conclusions about. ⇨ **fixed effects model**. ⇔ **random effect**

fixed effects model   a statistical model that assumes we wish to make inferences about the particular **levels** of a **factor** used in the study, and no others. This is particularly relevant when including study centre as a factor in the analysis: do we wish our results to be applicable only to those centres that took part in the study, or do we wish to consider those centres to be a random selection of all the centres that might have taken part so that the results can be applied to all possible centres? The fixed effects approach assumes the first case. ⇔ **random effects model**

fixed sample size design   a design that determines the number of subjects to be recruited before the study starts and does not allow the number to be changed. This is the most common type of approach to determining how many subjects should be in a study. ⇨ **group sequential design, interim analysis, sequential design**

flat file   a computer datafile that can be thought of as like a **matrix**, usually with each row representing one subject and each column representing one variable. ⇔ **hierarchical database**

floor effect   an **asymptote** that is a lower limit. Often zero will be that lower limit. ⇔ **ceiling effect**

floppy disk   a form of computer **disk** that is easily portable and is intended to be slotted in or out of a computer rather than being a permanent fixture (as in **hard disk**)

flow diagram   a diagram showing a series of activities occurring across time (Figure 10)

flowchart   ≈ **flow diagram**

follow-up   the process of collecting data after some activity has taken place. This often simply means gathering data after subjects have been **randomised**, or it may mean collecting data after treatment has been stopped to monitor **safety** or **relapse** of symptoms

follow-up data   data that are collected as a result of **follow-up**

follow-up period   the time during which **follow-up** occurs. This may simply be the time that patients are in a study from randomisation until their last visit

follow-up visit   any **visit** during a **follow-up period** of a study

for cause audit   an **audit** that is carried out because of some suspicion of poor quality work or of fraud. ⇔ **no cause audit**

form   ≈ **case record form**

formulation   the way in which a product is manufactured and presented. Examples include **tablets, capsules, injections**. ⇨ **product**

FORTRAN   a very powerful but quite old computer programming language. ⇨ **BASIC, Visual Basic, C, C++**

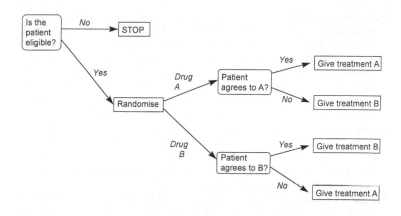

Figure 10 Flow diagram. The sequence of events to follow in Zelen's randomised consent design, seeking consent in conjunction with randomisation

forward selection a method of arriving at a **regression model** when several possible **covariates** might be included. The method begins by selecting the variable that makes the greatest contribution to reducing the **residual variance** (subject to some minimum criterion) and putting this in the model. Then the variable giving the next greatest reduction in variance (again, subject to a minimum criterion) is found and included in the model. The process continues until either all the variables are in the model or no more meet the minimum criterion for being included. The minimum criterion is referred to as *F* **to enter**. ⇨ **all subsets regression, backward elimination, stepwise regression**

forward stepwise regression ≈ **forward selection**

fourfold table ≈ **two-by-two table**

fourth hurdle some **regulatory authorities**, in addition to requiring demonstration of quality, safety and efficacy, require evidence of additional value for money of a new product. This is called the 'fourth hurdle'. ⇨ **pharmacoeconomics**

frailty model a statistical model that assumes different individuals have different probabilities of being unobserved. The term is most often used with respect to **survival times** where it is expected that there will be some **censored data**. Survival models assume that the probability of

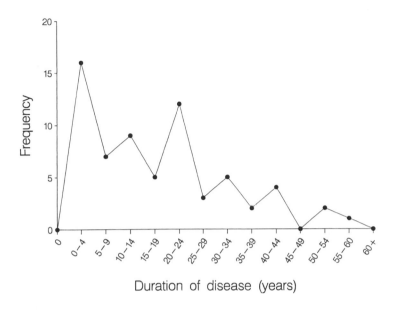

Duration of disease (years)

Figure 11 Frequency polygon. Distribution of the number of years a group of 87 patients had suffered from eczema. Only the outline of the histogram is plotted

censoring is the same for every subject but frailty does not make that assumption

frame ≈ **sampling frame**

fraud the act of intentional and dishonest deception. ⇨ **fabricated data**

fraudulent data ≈ **fabricated data**

free combination therapy a mixture of two (or more) drugs that are intended to be taken together but which are not combined in one **formulation**. ⇔ **fixed combination therapy**

frequency the number of times a particular event occurs or a particular data value is observed. ⇔ **relative frequency**

frequency distribution the number of times each of several **events** occurs or the number of times each of many different data values occurs. ⇨ **frequency polygon, frequency table, histogram**

frequency polygon a diagram for representing a **frequency distribution**.

Table 6 Frequency table of extent of body surface area affected by eczema in 157 patients

	Frequency	Percentage	Cumulative frequency	Cumulative percentage
No involvement	42	26.8	42	26.8
<10%	48	30.6	90	57.3
10–29%	25	15.9	115	73.2
30–49%	14	8.9	129	82.2
50–69%	16	10.2	145	92.4
70–100%	12	7.6	157	100.0

Each of the data values is placed along the *x* **axis** and the number of times each occurs is plotted as a point on the *y* **axis**. These points are then joined to form a polygon (Figure 11). ⇔ **histogram**

frequency table a numerical summary of a **frequency distribution** showing the number of times each data value occurs. Sometimes this may be enhanced to also show the percentage of occurrences, the **cumulative frequency** and the cumulative percentage of occurrences. All of these features are shown in Table 6

frequentist inference an approach to data analysis that produces **estimates** of **parameters, confidence intervals** and **significance tests**. ⇨ **Bayesian inference, fiducial inference**

Friedman's test a **nonparametric significance test** for testing the **null hypothesis** that all of several treatments given to the same subjects have the same distribution of responses. Informally, this can be thought of as the nonparametric equivalent of **repeated measurements analysis of variance**

Friedman's two way analysis of variance ≈ **Friedman's test**

full analysis set ≈ **intention-to-treat population**

full model ≈ **saturated model**

fully compliant a subject who takes or uses all medication exactly as prescribed in the study **protocol**

function a mathematical equation

funnel plot a type of graph for plotting summary results from many different studies. It is used in **meta-analysis** and in **overviews** to help try to detect **publication bias** (Figure 12)

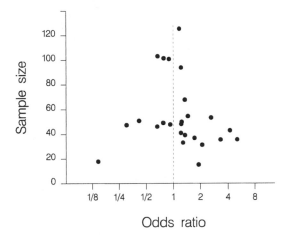

Figure 12 Funnel plot. Summary odds ratios from 25 studies comparing the efficacy of a certain class of antidepressant with placebo. If no publication bias existed, we would expect to see a 'funnel' shape. There is some suggestion here that some small negative studies may have been missed because of the lack of studies in the bottom centre of the plot

Galbraith plot ≈ **radial plot**

Gaussian curve ≈ **Normal distribution**

Gaussian distribution ≈ **Normal distribution**

Gehan's design a design, typically in **Phase II** cancer studies, where no **control group** is used. The design initially recruits a small number of patients: if overwhelming evidence in favour or against **efficacy** is seen then the study stops. If the evidence is not conclusive either way, further patients are recruited in order to obtain a reasonable estimate of the treatment **response rate**

Gehan's generalised Wilcoxon test a **nonparametric** statistical **significance test** for comparing two **survival distributions**. ⇨ **Cox's proportional hazards model, log rank test**

gel a **vehicle** for delivering **topical** treatments. Similar to **cream** but more solid. ⇨ **lotion, ointment**

gender synonym for sex

general linear model ≈ **linear model**

generalisability the extent to which conclusions can be applied to a wide **population**. ⇨ **external validity**

generaliseable conclusions that have wide **generalisability**

generalised additive model a method of producing models, similar to **generalised linear models**, that predict an **outcome variable** from several **independent variables**. In this case, the **link function** is a complex function of the data, rather than a theoretical link function, such as the **Normal distribution** or **logistic function**

generalised estimating equations an extension to **linear models** particularly useful for modelling **repeated measurements** and further particularly suited to **binary data** and **Poisson data**

generalised linear model an extension to **linear models** where a **link function** is introduced. This link function is a function of the **response variable** and, instead of modelling the response variable directly, the link function is modelled as a **linear function** of the **independent variables**

Table *x*   Mean Systolic Blood Pressure at Baseline
and After Four Weeks of Treatment

	Treatment *A*	Treatment *B*
Baseline		
Mean	xxx.x	xxx.x
Std deviation	xx.x	xx.x
Minimum	xxx	xxx
Maximum	xxx	xxx
n	xx	xx
After treatment		
Mean	xxx.x	xxx.x
Std deviation	xx.x	xx.x
Minimum	xxx	xxx
Maximum	xxx	xxx
n	xx	xx

Figure 13   Ghost table. All of the row headings and column headings
have been drafted out so it is clear what the table will look like when the
data are available. In this example, the number of decimal places and
significant digits for each of the numerical values have also been indicated

generic   the fundamental, original, form. Often used to refer to drug
names as **generic name**, in contrast to **trade name**

generic name   the name that the original manufacturer or developer gives
to a drug. ⇔ **trade name**

genetics   the study and description of genes and DNA

Genie score   a way of summarising **multivariate data** (usually used for
laboratory data) The greater the score, the more deviation there is in a
subject's (laboratory) data from the relevant **reference ranges**

geometric mean   a measure of **central tendency**, particularly useful for
highly **skewed data.** It is calculated as the *n*th root of the product of *n*
numbers or, alternatively, as the antilog of the mean of the logarithms of
all the numbers. ⇨ **harmonic mean**

ghost report   a draft of a report that contains no results but has all the
section headings and some of the introductory text included. The

intention of a ghost report is to be able to produce a final report as quickly as possible after the data become available. A ghost report may also contain **ghost tables**

ghost table   the layout of a **table** indicating row and column headings but without any data (Figure 13). ⇨ **ghost report**

Gini coefficient   a measure of **variation** most often used in describing income or salaries. Hence it has uses in **pharmacoeconomics**

glossary   a list of specialist terms referred to in a document, with their definitions

goal   a target such as the goal for the number of subjects to be recruited to a study

gold standard   a **diagnostic test** that is guaranteed to give the correct **diagnosis**. Also used to refer to a treatment that is widely recognised as the best available

golden rule   informal term for 'most important rule'

Good Clinical Practice (GCP)   a set of principles and guidelines to ensure high quality and high ethical standards in clinical research. ⇨ **Good Laboratory Practice**, **Good Manufacturing Practice**

Good Distribution Practice (GDP)   a set of guidelines to ensure high quality standards in warehouse storage and distribution work

Good Laboratory Practice (GLP)   a set of guidelines to ensure high quality standards in laboratory work. ⇨ **Good Clinical Practice**, **Good Manufacturing Practice**

Good Manufacturing Practice (GMP)   a set of guidelines to ensure high quality standards in manufacturing. ⇨ **Good Clinical Practice**, **Good Laboratory Practice**

Good Regulatory Practice (GRP)   a set of guidelines to ensure high quality standards in regulatory affairs work

Good Statistical Practice (GSP)   a set of guidelines to ensure high quality standards in statistical work

goodness of fit   a measure of agreement between a set of **observed data** and a **model** that has been fitted to those data

goodness of fit test   a statistical **significance test** to compare whether one **model** fits data better than an alternative model

Graeco-Latin square   a form of **Latin square** that balances for three sources of variation. ⇨ **Youden square**

grand mean   the mean of a set of numerical observations, regardless of which group (treatment group or other form of group) those data relate to. ⇨ **grand total**

grand total   the total of a set of numerical observations regardless of which group (treatment group or other form of group) those data relate to. This equates to the **grand mean** multiplied by the number of observations

graph   a pictorial representation of data plotted on an *x* axis and *y* axis, and sometimes on a *z* axis too. ⇨ **scatter plot**

graphic   a general term for diagrams, graphs, sketches, etc.

Greenhouse–Geisser correction   an adjustment made to the **degrees of freedom** in an *F* **test** of **within subjects effects** in **repeated measurements analysis of variance**. It is assumed that the pattern of **correlation** is constant over time and this adjustment is required if the assumption is not valid. ⇨ **Huynh–Feldt correction**

group   one of the **strata** in **stratified data**. The term is frequently used to refer to subsets such as the **treatment group** or the **placebo group** (those treated with active treatment or those treated with placebo, respectively). It can be used to refer to other strata such as the males or females, the 'high risk group', 'low risk group', etc.

group data   the subset of an entire set of data that relates to only one group. For example, all the data from subjects treated with placebo or all the data from female subjects

group ethics   ≈ **collective ethics**

group matching   usually in matching we refer to **matched pairs**. However, with group matching we imply that overall, two (or more) groups of subjects are typically quite similar in terms of their **demographic data**, **disease profile**, etc.

group randomisation   ≈ **cluster randomisation**

group sequential analysis   special types of analysis that are appropriate for **group sequential studies**

group sequential design   a form of **sequential design** where **interim analyses** are carried out after a number of subjects have been recruited into a study. Usually only two or three analyses would be planned into such a study after either half the subjects or one third and two thirds of the subjects have completed the study. ⇨ **O'Brien and Flemming rule, Pocock rule**

group sequential study   a study designed as a **group sequential design**

group sequential test   a statistical **significance test** carried out in **group sequential designs**

grouped data   ≈ **categorical data**

growth curve   a graph that traditionally plots the progress of some feature of growth over time. Growth could be measured by height or weight. The term now has a broader use to include any variable that systematically changes (usually increases) over time

guardian   ≈ legal guardian

guesstimate   an informal term to describe a result that is largely a guess but that supposedly has some data used to help form that guess. It is not an **estimate** in the formal sense of the word but it is supposed to be better than a guess based on no knowledge (or data) at all

guideline   a set of suggested rules but ones that are not enforceable by any laws. In practice, one would be foolish to ignore many of the regulatory guidelines that have been written. ⇨ **Good Clinical Practice**

guinea pig   a subject who is part of a study may be referred to as a guinea pig. The term is used disparagingly by those who do not approve of the particular study involved (or who do not approve of research on humans generally); or it is used light heartedly. Given these two extremes of meaning, it is a term best avoided

Guttman scale   a method of combining answers to individual questions to arrive at an overall score (sometimes called a composite score). Each question may be **weighted** differently, so it is not simply the sum of the individual question responses

H

H_0 symbol for **null hypothesis**

H_1 usual symbol for **alternative hypothesis**

H_a alternative (less often used) symbol for **alternative hypothesis**

haematology   the study of the makeup of blood. Usually used in the context of **laboratory data** to refer to such parameters as platelets, red blood cell counts, etc. ⇔ **biochemistry**

half life   the amount of time that a radioactive substance takes to decay to half its original quantity or a drug takes to halve its **concentration** in the body. In many situations, four half lives might be considered a reasonable time to reduce the original concentration to a minimal quantity (four half lives being $\frac{1}{16}$ of the original amount)

halo effect   an informal term to describe **psychosomatic effects** that often occur when patients believe that the doctor will be able to give them benefit. ⇨ **placebo effect**

handbook   a book of instructions for using a machine, for running a study or for general work practices

haphazard   unpredictable but not in the highly controlled sense of **random**

haphazard sample   a sample of people (or items). The members of the sample are not chosen for any particular reasons, just as they happen to present themselves. Haphazard samples often display various patterns that would not be seen in a truly **random sample**. ⇨ **convenience sample**

haphazard treatment assignment   a method of assigning treatments to subjects that is not **controlled** or predictable. Like **haphazard samples**, haphazard treatment assignment often displays various patterns that would not be seen in truly **random assignment**

hard data   ≈ **objective data**

hard disk   a form of computer disk that usually resides inside the computer and is not intended to be moved between different computers. They have much larger capacity than **floppy disks** or **diskettes**

hard endpoint   ≈ **objective endpoint**

hard measurement   ≈ **objective measurement**

hard outcome   a response to an intervention that can be measured using **objective data**. ⇔ **soft outcome**

hardware   the mechanical, electrical and electronic components of a computer such as the screen, the processor, disk drives, keyboard, etc. ⇨ **software**

harmonic mean   a measure of **central tendency** used for **skewed data**. It is calculated using the reciprocals of the data, namely $H = \{\frac{1}{n}\Sigma(1/x_i)\}^{-1}$. ⇨ **geometric mean**

Hawthorne effect   the response that is often seen in subjects taking part in a study and produced simply because they know that they are being observed; however, it is not a true effect of any intervention. The strict definition given for **effect** is particularly important to note. ⇨ **placebo effect**

hazard   ≈ **hazard function**

hazard function   in **survival analysis** the probability of a given **event** (such as death) occurring at each instant in time, given that the event has not already happened. ⇨ **Cox's proportional hazards model**

hazard rate   the **hazard function** at any particular point in time

hazard ratio   the ratio of two **hazard rates** or of two **hazard functions**, either at a particular point in time or averaged over a long period. ⇨ **Cox's proportional hazards model**

health   the general state of wellbeing or lack of wellbeing in an individual or a group of individuals

health economics   ≈ **pharmacoeconomics**

health services research   research into the provision of health care, including aspects of cost, need, resources, supply and outcome. Strongly linked with **pharmacoeconomics**

healthy subject   ≈ **healthy volunteer**

healthy volunteer   a subject who volunteers to take part in a study but who does not have any significant disease. Such subjects often participate in **Phase I studies**. Note that all subjects who take part in clinical trials should do so voluntarily; for this reason, the term should not be abbreviated simply to 'volunteer' (although it often is). ⇔ **patient**

healthy worker effect   a form of **volunteer bias**. Subjects who have employment (those that are workers) tend to be healthier, on average, than the general population (which includes those who do not work, through choice, old age, disability, etc.)

Heisenberg effect   a term from physics that says that the act of observing and measuring a process affects that process so that absolute effects are impossible to measure. This is one of the reasons why we need **comparison groups** in studies. ⇨ **Hawthorne effect**

Helmert contrasts   a particular type of **contrast** where each **level of a factor** is compared with the mean of all other levels of that factor. For example, if three ethnic groups are represented in a study the **response variable** could be investigated to see if it is affected by ethnic group. The mean response in ethnic group 1 would be compared with the mean of the combined data from ethnic groups 2 and 3; ethnic group 2 would be compared with the mean of the combined data from ethnic groups 1 and 3; ethnic group 3 would be compared with the mean of the combined data from ethnic groups 1 and 2. ⇨ **analysis of variance, multiple comparisons**

hepatic metabolism   **metabolism** (of product) through the liver. ⇨ **renal metabolism, pharmacokinetics**

heterogeneous   a term used to mean that the variation of a measurement within a group is different from the variation of that same measurement within other groups. ⇔ **heteroscedastic, homogeneous**

heteroscedastic   unequal **variances** of data values of the same variable. For example, the variation in the measurement of a person's age usually changes with age; age of newborns may be measured in hours or days, age of infants in months, adults in years. ⇔ **heterogeneous**

heuristic   using intuition and judgement

hierarchical   nested, meaning built up in layers

hierarchical database   a computer database that has several levels of data. For example, the highest level may be the subject level recording basic **demographic data** for each subject. For each subject, other levels may contain data on the diseases they have and, for each disease, the treatments they have been given. ⇔ **flat file**

hierarchical models   two statistical **models** for the same data but one has extra **covariates** that are not included in the other

high level term   a classification of signs, symptoms and diseases (particularly used in **MedDRA**) giving a coding that is less detailed than the **preferred term** but is more detailed than the **system organ class**

high order interaction   a general term used to refer to an **interaction** (in the statistical sense) that is not a **two factor interaction** but involves at least three **factors**

highest density region   in **Bayesian** statistics the middle region of a **posterior distribution** used for determining **interval estimates**. ⇨ **credible interval, confidence interval**

highest posterior density   a method in **Bayesian** statistics for determining a **point estimate** of a **parameter**

high–low graph   a graph for plotting one **continuous variable** (usually on

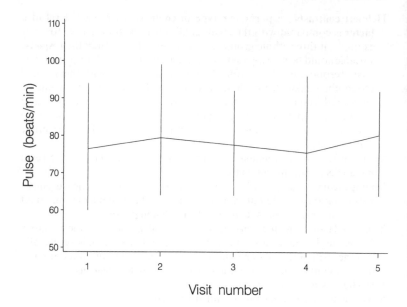

Figure 14 High–low graph. The mean pulse rate in 100 patients with ischaemic heart disease is plotted at each of five visits. Additionally, the minimum and maximum values at each visit are plotted. Note that the patient with highest pulse at visit 1 is not necessarily the one with the highest pulse at any other visit

the *y* **axis**) against one **categorical variable** (usually on the *x* **axis**). It shows the **mean** and/or **median** and the minimum and maximum values of the continuous variable for each value of the categorical variable (Figure 14)

hinge ≈ **quartile**

Hippocratic oath a promise to act to certain high ethical and medical standards. Traditionally it is thought of as being sworn by all doctors when they qualify but this is not actually the case

histogram a graphical method for plotting a **frequency distribution** similar to a **bar chart**. Whilst a bar chart is typically used for categorical data, a histogram is more usually used for continuous data (Figure 15)

historical control a **control group** that has not been **randomised** but consists of patents treated in the past. ⇔ **concurrent control**

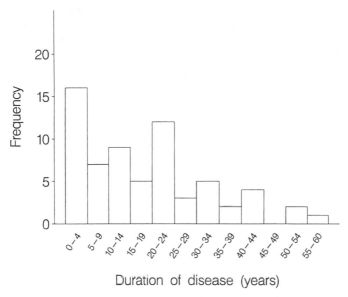

Figure 15 Histogram. Distribution of the number of years a group of 87 patients had suffered from eczema

historical control group a **comparator group** that has not been assigned by **randomisation** but which consists of patients treated in the past (sometimes patients who were not treated). This is a much less desirable method of making comparisons and is prone to many forms of unpredictable **bias** but it is a much easier source of comparisons than setting up a randomised study

hold constant when analysing data and producing **adjusted estimates** (by **analysis of covariance** or some other method) it is convenient to think of the result as what would have been observed had a particular **covariate** taken the same value for every subject. This is sometimes described as that covariate being 'held constant'

home visit in many studies, subjects are seen at hospital, at their own general practice or at some other kind of health centre. Particularly in **community studies**, a home visit is when a nurse, doctor or other health professional assesses the subject in their own home

homeopath one who practices **homeopathy**

homeopathy a treatment regimen that involves exposing patients to trace amounts of a chemical that would, in large enough doses in healthy people, produce symptoms of the disease that is being treated

homogeneous the variation of a measurement within a group being similar to the variation of that same measurement within other groups. ⇔ **heterogeneous, homoscedastic**

homoscedastic equal variances of data values of the same variable. For example, the variation in the measurement of a person's weight would not be expected to vary between different treatment centres (even though the mean might vary considerably). ⇔ **heteroscedastic, homogeneous**

hot deck a method of **imputing** for **missing data** based on other non-missing data

Hotelling's T^2 test a statistical **significance test** for comparing the means of two **multivariate distributions**. It could be used, for example, when a subject's 'size' is measured on three variables: height, weight, and head circumference. Rather than three separate t **tests**, the Hotelling test compares 'size' rather than separately comparing height, weight, and head circumference

Huynh–Feldt correction an alternative to the **Greenhouse–Geisser correction** in **repeated measurements analysis of variance**

hypothesis a statement for which good evidence may not exist but which is to be the subject of an experiment. A common example in clinical trials would be that 'Drug A shows an effect identical to that of placebo'. This is clearly a statement; it may be true or false; it can be tested in an appropriately designed experiment. ⇨ **alternative hypothesis, null hypothesis**

hypothesis generating study a study that is not intended to answer specific questions but rather to produce data that can be looked at in various ways to suggest interesting questions (or **hypotheses)** to be researched in subsequent experiments. ⇔ **definitive study**. Many studies may be run with the intention of answering a small number of hypotheses and to generate further ideas

hypothesis test a statistical process to determine the **strength of evidence** in favour of, or against, a particular **hypothesis**. There are many types of hypothesis test for use in different situations and for addressing different types of question. ⇨ **nonparametric test, parametric test**; and, for example, **chi-squared test, F test, Mann–Whitney U test, t test, P-value**

hypothesis testing the process of using a statistical **hypothesis test** to test a **null hypothesis**

hypothetical population   a **population** that cannot be completely defined (it would not be possible to list the names of all the individuals in that population, for example) but that can be considered to exist for practical purposes. ⇨ **infinite**. ⇔ **finite population**

I

iatrogenic   describing a condition caused by the treatment given for another disease. Obvious examples are **adverse reactions**

id ≈ **subject id**

identification number ≈ **subject identification number**

ignorable missing data   data values that, despite being missing, do not introduce any **bias** into the analysis and results of a study. ⇨ **missing completely at random, missing at random**. ⇔ **informative missing data, nonignorable missing data**

ignorable missingness   the process that produces **ignorable missing data**

ignorant prior   in **Bayesian** statistics, a **prior distribution** that gives no (or very little) information. ⇨ **improper prior, reference prior**

imbalance   lack of **balance** or not balanced

immune   not susceptible to a disease

immune system   those parts of the body, particularly antibodies, that help to protect or fight against infection

immunise   to make someone **immune** to a particular disease. This may occur either naturally or artificially by inoculation

impartial witness   someone who observes an event (usually that of giving **informed consent**) but who has no involvement with the study

improper prior   in **Bayesian** statistics, this is a **prior distribution** that is not a valid **probability distribution** but which can still be used as if it were. In general it states that our prior belief about a parameter is that it lies somewhere between minus infinity and plus infinity. As this does not tell us much about the parameter, it is sometimes called an **ignorant prior**. ⇨ **reference prior**

imputation   the process of **imputing**

impute   to fill in data values (usually **missing data**) with values that are thought to be sensible. There are several ways of doing this; many make valid assumptions, many make very questionable assumptions. Some methods rely on calculations based on the remaining data, some rely on intuition and guesstimates. The most common example is probably the concept of **last observation carried forward**

in vitro   in a test tube (or similar). ⇔ *in vivo*

in vivo   in living tissue. ⇔ *in vitro*

inactive control   a **placebo**. Use of the term 'control' indicates that some intervention (even if only placebo) is implied. The term would not generally be used to refer to a control group that received no treatment at all

incidence   the number of new cases (of a disease) that occur in a specified period of time. ⇔ **prevalence**

incidence rate   the number of new cases of a disease in a period of time, divided by the number of subjects at risk of the disease. ⇔ **prevalence rate**

incident   ≈ **event**

inclusion criteria   the requirements that a subject must fulfil to be allowed to enter a study. These are usually devised to ensure that the subject has the appropriate disease and that he or she is the type of subject that the researchers wish to study. Inclusion criteria should not simply be the opposites of the **exclusion criteria**

incomplete block   a **block** of treatment (or **treatment sequences**) that does not contain all of the possible treatments (or treatment sequences) to which subjects in the study may be randomised. ⇔ **complete block**

incomplete block design   a study that uses **incomplete blocks** of treatment. Although each block will necessarily be unbalanced (which may not be desirable), the study as a whole can still be **balanced**, as in a **balanced incomplete block design**

incomplete crossover design   a **crossover design** where not all subjects receive all of the possible treatments

incomplete crossover study   a study that is designed as an **incomplete crossover design**

incomplete factorial design   a **factorial design** where not all combinations of the possible treatments are used

incomplete factorial study   a study that is designed as an **incomplete factorial design**

increment   an increase in value (commonly the dose of a drug or the draft number of a protocol). ⇔ **decrement**

incubation period   the time between **exposure** to an **infection** and the appearance of clinical signs. ⇨ **sojourn period**

independent   if knowledge of one event or variable gives us no information (or even clues) about another event then the two events are said to be independent of each other. ⇨ **correlation**

independent contrasts   two (or more) **contrasts** that are **independent** of each other. If we were to compare the mean responses in three treatment groups (A, B and C), there are several possible contrasts that we could make. The simplest would be to compare each pair of treatments: mean(A) − mean(B), mean(A) − mean(C), and mean(B) − mean(C); however, if we know that A is greater than B, and that B is greater than C, then we immediately know that A must be greater than C. So these three contrasts are not independent of each other

independent ethics committee ≈ **research ethics committee**

independent groups   groups of subjects that are **independent** of each other. For example, a **parallel group design** uses independent groups but a **crossover design** does not

independent identically distributed (iid)   a term used to describe values of a **random variable** that are **independent** of each other but which all come from the same underlying **probability distribution**. In a **random sample** of women, shoe sizes might all be independent, and all from the same distribution; if the sample contained men and women then, although the shoe sizes may all be independent, there might be two underlying distributions (larger shoes for men than women)

independent random variable ≈ **independent variable**

independent samples ≈ **independent groups**

independent samples *t* test   a statistical **significance test** for testing the **null hypothesis** that the means of two populations are equal. ⇔ **paired *t* test**

independent variable   another term for a **covariate** in a **regression model**. Note, confusingly, that several so-called independent variables may not be **independent** of each other, nor of the **response variable** (or **dependent variable**). In a regression model the response variable may depend on the independent variables but the independent variables are not dependent on the response variable. For example, blood pressure may be partially predicted from knowing a subject's age, height, weight, etc.: these variables would be said to be the independent variables, whilst blood pressure is the dependent variable

index case   a **case** (as in **case-control study**)

index group   all of the **cases** in a **case-control study**

indexed file   a term that might be obvious in keeping paper files but is more relevant in computer databases. It is a collection (a file) of data that has an index which allows **direct access** to the required items

indication   the reason for using a product or other intervention. Synonym for disease

indicator variable   in computing terms a variable that is a **binary variable**. Often a set of indicator variables may exist to describe the values of one categorical variable. If a subject is randomised to receive one of three treatments, two indicator variables can be set up: the first takes the value 1 (and the second 0) if the subject is randomised to Treatment A; the first variable takes the value 0 and the second 1 if the subject is randomised to Treatment B; otherwise, both variables are set to 0, indicating that the subject must have been randomised to Treatment C

indirect contact   the contact of one person with another through a third party. Particularly relevant with infectious diseases, where the infection may initially be passed to someone in **direct contact** with the source of infection but these people may then pass the infection on further

indirect cost   in **pharmacoeconomics** a cost incurred because someone has a certain disease, but not the **direct cost** of treating the patient. Loss of earnings and social security payments are often considered indirect costs

individual   relating to a particular item (often, but not necessarily, a person)

individual ethics   ethical behaviour that focuses on benefit to an individual rather than benefit to society. ⇔ **collective ethics**

individual matching   finding **cases** and **controls** that have similar **demographic data** and/or **disease profiles**. For each case, one or more similar controls is sought for comparison. ⇔ **group matching**

individual variation   variation in measurements of individuals, rather than of groups. ⇨ **within subjects variation**. ⇔ **between subjects variation**

induce   to draw a conclusion or a generalisation from specific examples of data. ⇔ **deduce**

induction   ≈ **induce**

inductive inference   the process of drawing conclusions by **induction**. ⇔ **deductive inference**

inductive reasoning   a less strong term than **inductive inference**

inequality   a statement which says that two things are not equal. Sometimes there may be sufficient information to know that one item or quantity is larger or smaller than another; otherwise 'not equal' is all that can be said

inert   having no (biological) action. **Placebos** are often considered as being inert

infection   the implantation and growth of an organism

infectious   a disease that can be passed on via **direct contact** or **indirect contact** with other people

infer   ≈ **deduce**

inference   a conclusion drawn based on data and reasoning

inferential statistics   the branch of statistical methods concerned with drawing conclusions from data, typically by use of statistical **significance**

testing. ⇔ **descriptive statistics**

infinite   without bounds. In numerical terms, a number larger than any other can be. ⇔ **finite**. ⇨ **minus infinity, plus infinity**

infinite population   a **population** (which must be a **hypothetical population**) that contains an **infinite** number of individuals. For the purposes of statistical methods used in clinical trials, most populations are assumed to be infinite. ⇔ **finite population**

influence   to contribute substantially to a decision or conclusion

influential observation   a data point that has a lot of **influence** on a statistical model. Some **outliers** can be very influential observations but this is not always the case

informatics   the science of handling and processing **information** (usually in the form of data)

information   a term encompassing data but rather broader. Some say that the value of analysing data is to turn it into information

informative censoring   **censored data** where the process of censoring tells us something about the state of a subject. If censoring is **random** then we know only that data are censored; if subjects withdraw from a study because they are too unwell to attend the clinic, or because they are free of any **symptoms**, then we may have censored data but in both cases there is information (negative or positive) in the censoring. ⇨ **informative missing data**. ⇔ **noninformative censoring**

informative missing data   **missing data** where the reason that the data are missing tells us something about the state of a subject. ⇨ **informative censoring**. ⇔ **noninformative missing data**

informative prior   in **Bayesian** statistics any form of **prior distribution** that is not a **reference prior**. ⇨ **proper prior**

informed consent   the practice of explaining to subjects and informing them about the purpose of a study and seeking their agreement to participate on a voluntary basis. ⇨ **Declaration of Helsinki, ethics, research ethics committee**

injection   a method of delivering liquid medication into the body. ⇨ **subcutaneous, intramuscular, intravenous**

inlier   a data value that does not seem to be true, given all the other data values, usually because it is too typical or normal. This is an odd concept but can be applicable to **multivariate data**. ⇔ **outlier**

inotropic effect   the effect a drug has on the contraction of the heart. ⇨ **chronotrophic effect**

inpatient   a patient who is treated in hospital and usually stays in hospital overnight. ⇔ **outpatient**

input device   a device for getting data into a computer (this may simply be the keyboard or it may be a sophisticated blood analyser that feeds results directly into the computer)

input variable   another term for a **covariate** or **independent variable**

inspection   review of data and work practices by an independent reviewer (usually from a **regulatory authority**). ⇨ **audit**

instantaneous rate   the number of subjects who experience an event at a particular (small) point in time divided by the number who were at risk at that time. ⇨ **hazard function**

institution   in the context of clinical trials, a place where a study is undertaken; usually a hospital or similar establishment

institutional review board   ≈ **research ethics committee**

integer   a whole number (1, 2, 3, etc.), including negative numbers $(-1000, -6,$ etc.) but excluding any fractions or decimal numbers $(3\frac{1}{2}, -6.75,$ etc.)

integrity   honesty (when applied to a person); correctness (when applied to data). ⇦⇨ **fraud**

intelligence   broadness of understanding of, and the ability to solve, problems (practical or theoretical). It is a term that can refer both to humans and other animals. Note, therefore, that it is not the same as 'general knowledge'

intelligence test   a series of questions and problems to measure **intelligence**. Often referred to as IQ (intelligence quotient) tests

intention-to-treat   a term very similar to **analysis by randomised treatment**. It is a strategy for analysing study data which (in its simplest form) says that any subject randomised to treatment must be included in the analysis. This is not always easy, particularly in the presence of **missing data**. ⇦⇨ **per protocol analysis**

intention-to-treat analysis   ≈ **intention-to-treat**

intention-to-treat population   the subset of subjects recruited into a study who are included in the **intention-to-treat analysis**

interaction   the joint influence of two or more **independent variables** on a **response variable** that is not simply the sum of the individual influences

interaction effect   the difference in the size of the **effect** caused by two (or more) variables jointly, compared with the sum of the individual effects. For example, it is known that smoking and exposure to asbestos increase the risk of bronchial cancer. However, for smokers who are exposed to asbestos the risk is substantially higher than the sum of the individual risks. There is said to be an interaction between smoking and exposure to asbestos

intercept   in a **regression model** this is where the **regression line** crosses the *y* **axis** (which is the value of *y* when $x = 0$)

interim   part way through; before the entire (study) is completed

interim analysis   a formal statistical term indicating an analysis of data part way through a study, usually in the context of **group sequential studies.** ⇨ **sequential analysis**

interim look   a less formal term than **interim analysis**, used to describe a broader range of analyses of data part way through a study. These may include formal interim analyses or less formal summaries of data, without necessarily having broken the **blind**

interim report   this term may either be used informally to refer to a preliminary report (that is, not a **final report**) or more formally to mean the report of an **interim analysis**

interim result   the results of an **interim analysis** or **interim look**

interim review   a review of data part way through a study, often to check on data quality and completeness rather than in the sense of a formal **interim analysis**

intermediate variable   a variable that does not measure exactly what we want to know but which is a second-best alternative. ⇨ **surrogate**

internal consistency   in questionnaires this is used to describe the situation where different questions find the same information; a simple example is to record age and date of birth. Note that both responses may be consistent with each other but also that both may be wrong. A similar usage applies to results in a study report where, again, two sets of results may be based on incorrect data and so may be wrong—but if the two results agree with each other, then the report would be said to have internal consistency. ⇔ **external consistency**

internal pilot study   a form of **pilot study** where the data collected also form part of the data for the **main study**

internal validity   a statement or result that is **valid**, given a set of assumptions. If those assumptions are not correct then that statement or result may not be true

International Classification of Diseases   a coding system developed by the World Health Organization. Virtually every disease, illness, injury, etc. is given an alphanumeric code

Internet   a worldwide computerised communications network and source of information

interobserver agreement   the extent to which two (or more) people agree with each other when recording measurements. This can be important in **multicentre studies** where several investigators (possibly in several countries) are supposed to be assessing the same quantity. It is most often referred to in the context of **subjective data** rather than **objective data**. ⇔ **intraobserver agreement**

interobserver disagreement   the extent to which two (or more) people disagree with each other when recording measurements. More commonly referred to as **interobserver agreement**. ⇔ **intraobserver disagreement**

interobserver variation   the variation that almost always exists when more than one person measures the same quantity. This variation leads to **interobserver disagreement**. ⇔ **intraobserver variation**

interpolate   to calculate an unknown value between two known values. This is most often done in a **linear** way but more complex methods exist. The practice is often used when looking up conversion values in tables and finding that the exact value to be converted is not tabulated. The required result may be approximately determined by interpolation. ⇔ **extrapolate**

interquartile   between the **lower quartile** and the **upper quartile**

interquartile range   a measure of **variability**. The value of the **upper quartile**, minus the value of the **lower quartile**

interrater agreement   ≈ **interobserver agreement**

interrater disagreement   ≈ **interobserver disagreement**

interrater variation   ≈ **interobserver variation**

interrelate   ≈ **correlate**

intersect   the point on a graph where two curves cross each other. This also includes one curve crossing the x **axis** or y **axis** (≈ **intercept**)

interval   the range between two data values. ⇨ **class interval**

interval censored observation   data that are **censored** within a **time interval**. Generally, in censoring it is assumed that a subject's status is known until a particular time; thereafter it is censored. In interval censoring a subject may be seen once a week or once every three months, etc. and all that is known is that the subject's data became censored some time during that interval

interval data   ≈ **continuous data**

interval estimate   a range of values a **parameter** is likely to take that reflect the uncertainty and variability in measurements. The most common types of interval estimate are **standard errors, confidence intervals** and **credible intervals**. ⇔ **point estimate**

interval estimation   the process of determining an **interval estimate**. ⇔ **point estimation**

interval scale   ≈ **continuous scale**

interval variable   ≈ **continuous variable**

intervene   to take action, rather than to do nothing but observe

intervention   the action that is taken when one **intervenes**. In clinical trials the most common type of intervention is to give treatment (or **placebo**)

intervention study   an alternative term for a clinical trial. ⇔ **observational study**

interview   a series of questions that are asked of a subject. Interviews may be held face to face or as **telephone interviews**

interview study   a study carried out by interviewing subjects

intraclass correlation   **correlation** between two measurements of the same variables, in the same subjects, taken at two different times

intraclass correlation coefficient   the statistical measure of **intraclass correlation**. It is denoted r, as is the more usual **correlation coefficient**

intramuscular   into the muscle tissue. A method of delivering drugs by injection. ⇔ **intravenous, subcutaneous**

intranet   a type of **wide area network** that resembles the **Internet** but does not have unlimited public access

intraobserver agreement   the extent to which the same person can repeatedly make the same measurement. As with **interobserver agreement**, this is more relevant with **subjective data** than with **objective data**. ⇨ **reliability**

intraobserver disagreement   the extent of disagreement between **repeated measurements** of the same quantity taken by the same person. More usually referred to in the context of **intraobserver agreement**

intraobserver variation   the variation in a person's repeated measurements of the same quantity that results in **intraobserver disagreement**

intrarater agreement   ≈ **intraobserver agreement**

intrarater disagreement   ≈ **intraobserver disagreement**

intrarater variation   ≈ **intraobserver variation**

intravenous   into the blood stream. A method of delivering drugs by injection. ⇔ **intramuscular, subcutaneous**

intuitive   a decision reached by use of judgement and experience rather than based on data

invariant   lacking variation. The term is most often applied to a result that is found through analysis of data when that result holds for a variety of different methods of analysis and a variety of different assumptions about the data. It is the result that lacks variation, not the data

invasive   entering into the body, for example by needle to give an injection or to take blood or by an endoscope to take a biopsy

inventory   a list of items, typically of study materials, paperwork, medication, etc.

inverse correlation   this is usually used synonymously with **negative correlation** but more precisely means the **correlation** between one variable and the reciprocal of another variable

inverse logarithm   the reverse function of taking **logarithms**. For logarithms to base e, the inverse logarithm is the function e^x

inverse relationship   strictly, this term should be used when one variable changes in relationship to the **reciprocal** of another. However, it is often used when one variable increases as another (on average) decreases; this is more correctly called a **negative relationship**

investigate   to systematically observe and take measurements. Note that this does not necessarily encompass the term **experiment**

investigation   a particular variable or collection of variables that are observed

investigational centre $\approx$ **investigational site**

investigational device $\approx$ **medical device**

investigational device exemption (IDE)   an exemption similar to a **clinical trial exemption certificate**, issued to allow a **medical device** to be used in trials

investigational device study $\approx$ **medical device study**

investigational new drug (IND) application   an application to the US Food and Drug Administration (FDA) for permission to test a new drug in humans

investigational product   the product (usually) that is being researched. $\Rightarrow$ **experimental treatment**

investigational site   the place where the **clinical** work for a study takes place

investigator   the person who carries out the investigation. The term is very commonly used to refer to doctors who see subjects in a study and administer medication and record progress. Sometimes the investigator is not medically qualified—he or she may, for example, be a microbiologist in a study of an antibiotic

investigator initiated study   a study that is proposed and usually run and managed by an **investigator** rather than one that is proposed and managed by a pharmaceutical company. $\Leftrightarrow$ **sponsor initiated study**

investigator's brochure   a document prepared by a pharmaceutical company, for use by **investigators**, that summarises all the known relevant data (including safety, efficacy, pharmacodynamics, pharmacokinetics, etc.) regarding an **investigational product**

isometric graph   a graph that attempts to plot three dimensional data in two dimensions (Figure 16). $\Rightarrow$ **contour plot**, x **axis**, y **axis**, and particularly z **axis**

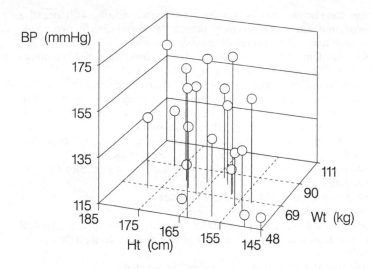

Figure 16 Isometric graph. A graph showing the relationship between height, weight and systolic blood pressure. In such graphs, the axis apparently going 'into' the page (in this case weight) is often referred to as the *z* axis

J

J shaped curve   a curve on a graph that resembles the shape of the letter J (for example **exponential growth**), or a reversal of it: ⌐ (**exponential decay**). The essential elements are that the curve is quite flat and then rises steeply, or that (in reverse form) it falls quickly and is then flat

J shaped distribution   a **distribution** that either has a large **peak** of values and then a long **tail** (a high degree of **positive skew**) or a long tail before a peak of values (high degree of **negative skew**)

jackknife   a statistical method of estimating **parameters** that helps to reduce **bias** in certain circumstances. The method calculates an estimate of the parameter of interest based on all the data except one observation; it then re-estimates the same parameter based on all the data except one other observation. The process is repeated until separate estimates have been calculated, each with the exclusion of one data point. These separate estimates are then combined

jackknife estimator   an **estimator** that uses the **jackknife** method

joint distribution   the **distribution** (either **frequency distribution** or **probability distribution**) of two (or more) **random variables**. To fully understand this distribution, we need to know the distribution of each of the variables separately and the **correlation** between them. ⇨ **bivariate distribution, multivariate distribution.** ⇔ **marginal distribution**

joint frequency distribution   see **joint distribution**

joint probability function   see **joint distribution**

Jonckheere–Terpstra test   a **nonparametric** statistical **significance test** for testing the **null hypothesis** of no **trend** in **ordered categorical data** between two or more groups

journal   a regularly published document containing academic research, reviews, etc.

judgement   use of intuition and experience instead of (or possibly in conjunction with) data

Kaplan–Meier curve a graph showing the **cumulative** probability of survival. ⇨ **Kaplan–Meier estimate**

Kaplan–Meier estimate a **nonparametric** estimate of the **cumulative** probability of **survival** for a set of data (that may include **censored observations**)

Kaplan–Meier product limit estimate ≈ **Kaplan–Meier estimate**

kappa (κ) coefficient an index (ranging from 0 to 1) of **interobserver agreement**

Kendall's tau (τ) a **nonparametric correlation coefficient**. ⇨ **Spearman's rho (ρ)**

keystroke error pressing the wrong key on a computer keyboard. The term is most often used in assessing quality of **data entry**, where the number of keystroke errors may be taken as a measure of the quality of the working practice. It is one of the major reasons for doing **double data entry**

kilobyte one thousand **bytes** of computer information. Desktop computers can typically store at least a million bytes (or 1000 kilobytes). ⇨ **megabyte**

kinetics ≈ **pharmacokinetics**

Kolmogorov–Smirnov test a **nonparametric** statistical **significance test** for testing the **null hypothesis** that the **location parameters** of two groups are equal. ⇨ **Kruskal–Wallis test, independent samples *t* test**

Kruskal–Wallis test a **nonparametric** statistical **significance test** for testing the **null hypothesis** that the **location parameters** of two or more groups are equal. ⇨ **Kolmogorov–Smirnov test, one way analysis of variance**

kurtosis a measure of how highly peaked a **distribution** is. Distributions that have steeper peaks than the **Normal distribution** are called leptokurtic; those that are flatter than the Normal distribution are called platykurtic. The Normal distribution is sometimes described as being mesokurtic (Figure 17)

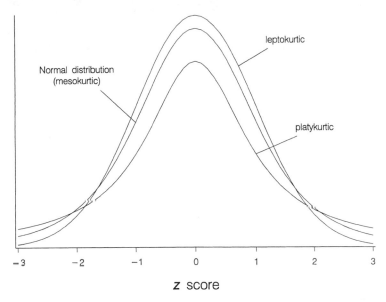

z score

Figure 17 Kurtosis. As distributions become more and more peaked they are called leptokurtic; as they become less peaked they are called platykurtic

L'Abbé plot a type of graph useful for plotting the results of many studies to assess how consistent they are with each other (Figure 18). ⇨ **meta-analysis, overview**

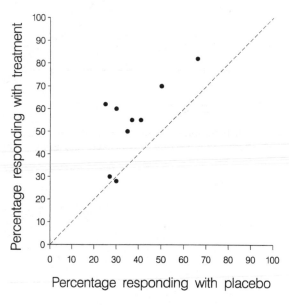

Figure 18 L'Abbé plot. Response rates from nine studies comparing an antipsychotic with placebo in obsessive–compulsive disorder. The diagonal is the line of equality: points above the line indicate that the response to treatment was better than that to placebo whilst points below the line (just one in this example) indicate a higher placebo response than treatment response

laboratory   a place where **investigations** and/or **experiments** are carried out. Traditionally, the laboratory was where experiments in chemistry or physics were done but the term is now used more broadly and may include, for example 'computer laboratory' or 'speech laboratory'

laboratory data   strictly any data that come from a **laboratory**. However, the term is usually used to refer to **biochemistry**, **haematology** and **urinalysis** data

lag   waiting behind. The term is used in computer programming and in statistical **time series** methods

landscape   a page that is wider than it is high, as in how (most) landscape pictures would be viewed. Landscape A4 paper is 297 mm wide and 210 mm high. ⇔ **portrait**

large sample method   ≈ **asymptotic method**

large scale trial   ≈ **megatrial**

Lasagna's law   the situation where the number of subjects eligible for a study apparently decreases when the study starts and increases again as soon as it ends. ⇨ **Münch's law**

last observation carried forward   a method sometimes used to analyse studies with **missing data**. Consider the situation where subjects are due to be seen at several visits (say, each month for six months), with the **endpoint** of the study being the six month assessment. If a subject withdraws from the study at month four, then we may use that month four data to replace the (missing) month six data. That is, we take the last actual observation and carry it forward to the end of the study. Various scenarios are illustrated in Table 7. ⇨ **intention-to-treat**

last visit analysis   ≈ **last observation carried forward**

last visit carried forward   ≈ **last observation carried forward**

latent period   ≈ **sojourn period**

Table 7   Individual subjects' heart rates (beats per minute) at four consecutive visits and each subject's value for the 'last observation carried forward'

Subject id	Visit 1 (baseline)	Visit 2 (4 weeks)	Visit 3 (8 weeks)	Visit 4 (12 weeks)	'Last observation'
1	98	99	94	89	89
2	80	72	Missing	Missing	72
3	83	83	80	81	81
4	95	90	Missing	95	95
5	110	88	80	Missing	80
6	88	Missing	Missing	Missing	88

Latin square   an experimental design that **balances** for two **sources of variation**. In clinical trials, the two sources are usually subject and time. The example in Table 8 shows how four treatments (A, B, C, and D) could be compared in four subjects in four time periods. The essential feature is that every treatment appears only once in every row (each subject) and once in every column (each time period). ⇨ **crossover study**, **Youden square**

Table 8   Latin square showing the sequence of four treatments (A, B, C and D) for four subjects in four periods

	Period			
	1	2	3	4
Subject 1	A	B	C	D
Subject 2	B	A	D	C
Subject 3	C	D	B	A
Subject 4	D	C	A	B

law   a set of rules. A relationship between a set of events and an **outcome**

law of averages   an informal term that reflects the fact that **probability distributions** exist and in particular reflects the belief that any particular outcome will eventually be observed if enough data are collected. It is distinctly different from the **law of large numbers** (or **central limit theorem**)

law of diminishing returns   ≈ **eighty–twenty rule**

law of large numbers   ≈ **central limit theorem**

lay person   someone who is not specifically trained in the subject being discussed but is nevertheless involved in discussing it. **Research ethics committees** will include lay members

lead time bias   a term often used in assessing **survival times** when the method of detecting **cases** improves with time. Patients apparently survive longer than they used to but this is not due to better treatment; rather it is due to earlier diagnosis. This could be an important problem in evaluating a **screening programme**. For example, even with no change in clinical practice, because cases may be detected earlier than without screening, the survival time from **diagnosis** will increase because diagnosis is occurring earlier in the life cycle of the disease

lead-in period   ≈ **run in period**

learning curve   a graph (rarely plotted, but frequently imagined) that plots time on the *x* **axis** and ability in a particular subject on the *y* **axis**. Sometimes such curves are **J shaped curves**, starting very flat and then rising steeply (suggesting it takes a long time before you can do anything, but then it all becomes clear); sometimes they start very steeply and then flatten off (suggesting it is easy to get started but learning the last few techniques becomes more and more difficult)

least significance difference test   ≈ **Tukey's least significant difference test**

least squares   a method of estimating **parameters** from data. It is based on choosing the value for that parameter that minimises the squared distance of each of the data values from the estimate of the parameter. ⇔ **maximum likelihood**

least squares estimate   an estimate of a **parameter** obtained by the method of **least squares**. ⇔ **maximum likelihood estimate**

least squares mean   the **estimated** mean of a variable obtained from an **analysis of variance** model or **analysis of covariance** model. It is the **adjusted** mean after adjusting for any other **factors** and **covariates** in the model

least squares method   any statistical method based on the principle of least squares. ⇔ **maximum likelihood method**

left censored   when measuring when an event occurs, the events that occurred before the study follow-up period (and so were not observed) are left censored. ⇔ **right censored**

left censored data   when the time of an event is known but the instant of **exposure** may be known only to be before a given time and the exact time is not known. Left censored data are much less common than **right censored data**. ⇨ **censored data**

left censored observation   ≈ **left censored data**

left skew   ≈ **negative skew**

left tail   the values in a **distribution** that are small (typically taken as meaning less than the **mode**)

legal guardian   someone who either permanently or temporarily is legally responsible for someone else's health and well being. ⇔ **next of kin**

lethal   will kill or extinguish life. ⇔ **fatal**

lethal dose   the dose of a drug that will kill an individual

lethal median dose (LD_{50})   the dose of a drug that will kill half of the subjects exposed to it

level of a factor   one of the different values that a **factor** (a **categorical variable**) can take. For example, the factor gender usually has two levels: male and female

level of blinding   whether a study is **open label**, **single blind**, **double blind**, **triple blind**, etc.

level of measurement   the degree of detail with which measurements are recorded. In general, the different levels (in descending order of detail) are **continuous data**, **ordinal data**, **categorical data** and **binary data**

Table 9 Example of a life table. In this instance, the radix (which is merely a baseline number taken for convenience) is 10000

Age (years) x	Survivors at age x	Deaths between x and $x + 1$	Probability of dying between x and $x + 1$
0	10000	2325	0.02325
1	7675	957	0.01247
2	6718	471	0.00701
3	6247	260	0.00416
4	5987	155	0.00259
5	:	:	:
:	:	:	:
:	:	:	:
50	2971	77	0.00259
:	:	:	:
:	:	:	:
90	119	27	0.0227
100	2	0	–

level of significance   in statistical **significance tests** this is the *P*-value (strictly speaking, specified before the calculations are carried out) that will be needed in order to declare a result as **statistically significant**. The most common cutoff value is 0.05 but 0.01, 0.001, etc., may also be used. Note that the level of significance is not the calculated (or observed) *P*-value

life expectancy   the length of time that an individual (or group of individuals) is expected to live

life table   a tabulation used to summarise **life expectancy** and probabilities of survival or death at different ages (or at different times after exposure to an intervention). An example of a life table is shown in Table 9. ⇨ **survival analysis**

life table analysis   methods used to analyse **life tables**, and particularly to compare **survival curves** between different groups of individuals and to assess the importance of **prognostic factors** on the length of survival. One of the most common methods is **Cox's proportional hazards model**

life table method   ≈ **life table analysis**

lifetime   the time between birth and death

lifetime prevalence   the **prevalence** of a particular event when the period within which the prevalence is measured is a person's entire lifetime. ⇨ **period prevalence**

likelihood   the **probability** of a set of observed data values, assuming a particular **hypothesis** (which is generally that they come from a particular **probability distribution** with specified **parameters**). Note that this is not the same as the probability of a given probability distribution, given a set of data. A variety of statistical procedures for **significance testing** and **estimation** are based on methods that use likelihood

likelihood function $\approx$ **likelihood**

likelihood principle   methods of estimating **parameters** and **significance testing**, based on **likelihood functions**

likelihood ratio   the ratio of the **likelihood** of two different **hypotheses** based on the same set of data

likelihood ratio test statistic   a general form of statistical **significance test** based on the **likelihood ratio**. Simplistically, the **hypothesis** with the greater **likelihood** is more likely (*sic*) to be correct

likert scale   an **ordinal scale** where scores are assigned to the different categories in the style of (for example) 1 = condition worse, 2 = no change, 3 = slight improvement, 4 = marked improvement, 5 = condition cleared

limit $\approx$ **asymptote**

line extension   an addition to a range of **products** or the range of uses of a product. This may include alternative forms of **presentation** or new **indications** for use

linear   in a straight line. $\Rightarrow$ **curve**

linear combination   a combination of values that is gained by simple addition and subtraction of multiples of those values. It does not involve any multiplication or other **nonlinear** functions of the values. For example, $x + \frac{1}{2}y$ is a linear combination of x and y but $x \times y$ and x^y are not linear combinations

linear correlation $\approx$ **correlation**

linear estimator   an estimator that involves only **linear combinations** of data values

linear kinetics   describes the **pharmacokinetics** of a product when the rates of **absorption**, **distribution** and **elimination** are each proportional to the dose of drug

linear model   a statistical model (such as a **regression model**) that only has a **linear combination** of parameters

linear regression $\approx$ a **regression model** that is a **linear model**

linear transformation $\approx$ **linear combination**

linear trend   a steadily increasing (or decreasing) response when a **covariate** increases (or decreases). The trend is linear if, for a fixed

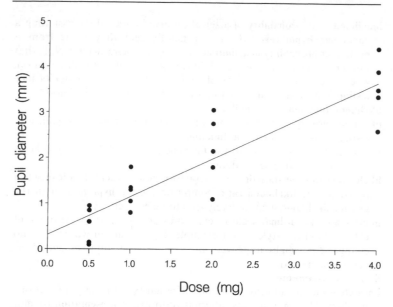

Figure 19 Linear trend. Pupil diameter after administration of a new test compound. One subject had five measurements taken one hour after receiving each of four separate doses of drug. Within the dose range used, the effect on pupil size seems to be linear

change in the covariate, there is a fixed size change in response (Figure 19). ⇨ **dose response relationship**

link function   a **transformation** of data values used to try to make a **nonlinear** curve be a **linear** one. Used extensively in **generalised linear models**

linkage   ≈ **record linkage**

literature review   a review of published studies and data relating to a particular topic. It is often the starting point for a new piece of research (to review the current and recent publications to find out what is known about a subject). It is also one of the first activities carried out in **meta-analysis** and **overviews**

ln   ≈ **log**$_e$

loading dose   a high dose of a drug that is initially given to quickly achieve a required **therapeutic** level. Thereafter, smaller doses (**maintenance**

doses) are often sufficient to keep the amount of drug in the body within the **therapeutic range**

local area network   a set of computers linked to each other to allow sharing of data and documents. The term 'local' is relative but tends to mean restricted to one site or building within a site. ⇔ **wide area network**. ⇨ **intranet**

local laboratory   a laboratory that is geographically close to where subjects are being investigated. ⇔ **central laboratory**

local research ethics committee   a **research ethics committee** that assesses studies to be carried out in local areas, typically with few centres. ⇔ **multicentre research ethics committee**

local server   ≈ **server**

location   a nonspecific term to describe the **central tendency** in a set of data

location parameter   the parameter used to describe **location** for any particular set of data. The most common location parameters are the **mean**, the **median** and the **mode**

lods   ≈ **log odds**

log   a systematic record of activities and actions. Also an abbreviation for **logarithm**

log odds $\log_e$ of the **odds** of an event occurring

log odds ratio $\log_e$ of the **odds ratio**. Many calculations concerning odds ratios are, in fact, carried out on the logarithm of the odds ratio and then transformed back to the odds ratio scale

log rank test   a statistical **significance test** for comparing the **survival times** of different groups of subjects

log transformation   the transformation of data values that is made by taking the **logarithm** of those data

$\log_{10}$ abbreviation for logarithm in base 10 units. ⇨ $\log_e$

logarithm   a mathematical function; the opposite function to the **exponential**

logarithmic transformation   ≈ **log transformation**

$\log_e$ abbreviation for logarithm in base e (e is a natural constant, approximately equal to 2.718). ⇨ $\log_{10}$

logistic curve   a **curve** that is the graph of the **logistic function** (Figure 20)

logistic function   a **transformation** of **binary data** that is used in **logistic regression**. Where the proportion of responses is denoted p, the transformation is $y = \log_e\{p/(1-p)\}$

logistic regression   **regression** where the **response variable** is **binary** and a **logistic transformation** has been used to help facilitate the mathematics in the statistical model. It is one form of **generalised linear model**

logistic transformation   ≈ **logistic function**

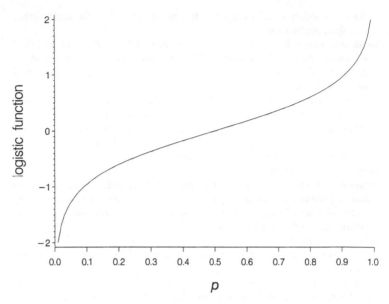

Figure 20   Logistic curve. The logistic function is defined for proportions (*p*) between 0 and 1. When $p = 0$, the logistic function equals minus infinity; when $p = 1$, the logistic functions equals plus infinity

logit   ≈ **logistic function**

logit model   ≈ **logistic regression**

log-linear model   a statistical model for analysing data that are in the form of a **count** of the number of observations that fall into each cell of a **contingency table**. It is one form of **generalised linear model**

lognormal distribution   the **probability distribution** of a variable such that the **logarithm** of that variable follows a **Normal distribution**

long term follow-up   usually restricted to observations on subjects after some **intervention** has taken place. The subjects may, or may not, be given medication during this time. 'Long term' is obviously open to interpretation but is generally considered to be at least six months. ⇨ **acute phase**

longitudinal   followed across time. ⇔ **cross-sectional**. ⇨ **cohort**

longitudinal analysis   the analysis of **longitudinal data**, usually with the specific intention of analysing changes with time. ⇨ **growth curve.** ⇔ **cross-sectional analysis**

longitudinal data   data that are repeatedly collected on the same subject across time. ⇨ **repeated measurements.** ⇔ **cross-sectional**

longitudinal study   a study that observes and measures the same subjects over a period of time. ⇔ **cross-sectional study**

loss   any negative effects of an **intervention.** A loss may be measured in cash, in years of life, in excess pain, etc. Sometimes a **negative loss** is referred to, meaning a gain

loss function   a function that combines several measures of **loss** (or possibly gain) to arrive at an overall figure for loss. The term 'loss function' is generally used when there is expected to be an overall loss (in the true negative sense); the term **utility function** is synonymous but tends to be used when there is expected to be a net gain (or **negative loss**)

loss to follow-up   the case when a subject is **lost to follow-up**

lost to follow up   a subject who supplies some data for a study but for whom after a certain time no more data are available. The term usually also implies that there is no known reason why the subject supplies no more data. ⇨ **censored observation, missing data**

lotion   a liquid used as a **vehicle** for delivering **topical** treatments. ⇨ **cream, gel, ointment**

lower quartile   the 25th **centile.** ⇨ **upper quartile, median**

main effect in **factorial studies**, the main effect of one **factor** is the size of the effect averaged over all levels of all other factors. ⇔ **interaction effect**

main study a term meaning **study** but useful to distinguish from **pilot study**

mainframe ≈ **mainframe computer**

mainframe computer a large computer. As technology progresses, the processing power and storage of small desktop computers is making the need for mainframe computers less and less. ⇨ **microcomputer, minicomputer, supercomputer**

maintenance dose the amount of drug that needs to be given to keep within the required **therapeutic range**. ⇔ **loading dose**

majority more than 50% (but not necessarily the **mode**). ⇔ **minority**

Mann–Whitney U test a **nonparametric significance test** for testing the **null hypothesis** that the **location parameter** (usually the **median**) is the same in each of two groups. ⇨ **independent samples t test**

Mantel–Haenszel estimate a method of estimating an **odds ratio** from a **stratified sample**

Mantel–Haenszel test a statistical **significance test** for testing the **null hypothesis** that the **Mantel–Haenszel estimate** of the **odds ratio** is equal to one

manual a set of instructions on how to use a machine or carry out a procedure

manuscript a written document sent to a publisher to be published (or to be considered for publication)

margin the edge. In **multivariate data**, each of the individual variables are sometimes referred to as the margins. See, for example, **marginal distribution**

margin of error ≈ **accuracy**

margin of safety ≈ **safety margin**

marginal see **marginal distribution, marginal mean**

marginal cost ≈ **per unit cost**

marginal distribution in **multivariate data**, the distribution of each of the variables, regardless of the other variables. Table 10 shows an example of **bivariate data.** ⇔ **conditional distribution, joint distribution**

Table 10 Joint distribution and marginal distributions of patients' and investigators' assessment of severity of disease in 202 patients. The marginal distributions are the row and column 'total' columns

Redness	Scaliness				
	Absent	Mild	Moderate	Severe	Total
Absent	1	1	0	0	2
Slight	25	12	3	0	40
Moderate	24	86	25	2	137
Severe	2	10	8	1	21
Very severe	0	1	1	0	2
Total	52	110	37	3	202

marginal effect   a loose term used to describe an effect that is quite small and that may not be real. ⇨ **marginally significant**

marginal mean   the mean of a **marginal distribution**

marginally significant   a loose term used to imply that a calculated *P*-**value** is very close to some arbitrary criterion for being called **statistically significant.** *P*-values of about 0.07 to 0.04 are often described as being marginally significant. ⇨ **marginal effect**

marker   ≈ **surrogate**

marketing authorisation   the authorisation given by a **regulatory authority** to a pharmaceutical company to market a product

mask   ≈ **blind**

match   to identify two (or more) subjects as having similar **demographic data** and/or disease severity (and other characteristics) such that one can serve as a **control** for another

matched control   a subject who has not been exposed to the intervention under study but who has **demographic data** and other exposure characteristics similar to those of one that has been exposed and who will be compared with that subject. ⇨ **case-control study**

matched design   a study where **matched pairs** are used

matched pair   two subjects who will be compared with each other or two measurements on the same subject that will be compared

matched pairs *t* test   ≈ **paired *t* test**

matched study   a study that is designed as a **matched design**

matched subjects   see **matched pair**

Table 11   Simple matrix of demographic data

Subject identification number	Age (years)	Gender	Race
1	37	Male	British
2	42	Male	British
3	18	Female	British
4	77	Male	Indian
5	45	Male	British
6	51	Female	American
7	52	Female	British

mathematical model   see **model**

mathematics   the science dealing with numbers, their uses and manipulation. ⇨ **statistics**

matrix   a rectangular (not necessarily square) array of mathematical elements. It could include numbers, **regression coefficients**, **parameter estimates**, etc. or simple raw data as shown in Table 11. Strictly it should have at least two rows and at least two columns: if it has only one row or only one column, it is called a **vector**

maximum   the largest of a set of values. ⇔ **minimum**

maximum likelihood   the largest value of the **likelihood function**. ⇨ **maximum likelihood method**

maximum likelihood estimate   the estimate of a **parameter** that is obtained by the **maximum likelihood method**

maximum likelihood method   a method of estimating **parameters**; the most likely value for the parameter (the best estimate) is the one that has the **maximum likelihood**. ⇔ **least squares method**

maximum tolerable dose   the maximum dose of a drug that a subject can take before inducing unacceptable **adverse reactions**

McNemar's test   a statistical **significance test** for testing the **null hypothesis** of no change in the proportion of subjects experiencing an event when each subject is assessed twice (under different conditions) and the data are in the form of **matched pairs**

mean   the sum of a set of numbers, divided by the number in the sample. A more formal term for the average

mean absolute deviation   ≈ **average absolute deviation**

mean square   the mean of a **sum of squares**

mean square error   the **variance** of an **estimator**. (Note that if the estimator is biased, then the mean square error is the sum of the variance of the estimator plus the square of the size of the bias)

meaningful difference $\approx$ **clinically significant difference**

measure   to determine the size or extent of a variable of interest

measured value $\approx$ **observed value**

measurement   the assessment and recording of a data value. This does not have to be restricted to **objective data**; the term is also used with reference to **subjective data**

measurement bias   a **bias** caused by the process of taking measurements. Examples include **digit preference** or measuring only values of a variable that fall within the capacity of the measuring instrument. $\Rightarrow$ **Hawthorne effect**

measurement error   an error made in measuring the value of a variable. The error may be because of lack of care in the measurement process or because of difficulty or judgement needed to measure the variable. Blood pressure, for example, is prone to measurement error, as are most types of **subjective data**

measurement scale   the type of scale that is used to measure a variable. Examples include **ordinal scale, continuous scale, categorical scale**, etc.

McdDRA   a dictionary of **adverse event** terms. McdDRA stands for Medical Dictionary for Drug Regulatory Affairs. $\Rightarrow$ **COSTART, WHO-ART**

median   the 50th **centile**. When a set of numbers is sorted into ascending order, there are as many values greater than the median as there are values smaller than the median. $\Rightarrow$ **lower quartile, upper quartile**

median dose   the dose of a drug that is estimated to produce a response in 50% of subjects

median life expectancy   the length of time that 50% of subjects are expected to live

medical   relating to **medicine**. $\Leftrightarrow$ **clinical**

medical device   a physical device used for medical treatment, such as a prosthesis or a heart pacemaker

medical device study   a study of the efficacy and/or safety of a **medical device**. This can encompass the comparison of more than one device or the comparison of a device and a pharmaceutical product

medical ethics   the branch of **ethics** that considers medicine, medical practice, medical care, etc. $\Rightarrow$ **Declaration of Helsinki**

medical history   the course of the health (including ill health) of a patient over time. This information can be used to help determine a diagnosis and predict a prognosis

medical judgement   a judgement (about diagnosis, treatment, prognosis, etc.) made by a physician

medical record   the notes and documents that describe a subject's medical history

medical study   a study of the efficacy and/or safety of one or more medicines. It is a more specific term than clinical trial

medical treatment   treatment administered to a patient. The type of treatment can be very broad but generally excludes **surgical treatment**

medical trial   ≈ **clinical trial**

medically important difference   ≈ **clinically significant difference**

medicine   the science and practice of prevention, diagnosis and treatment of disease

megabyte   a unit of space for storing information on a computer. Equivalent to one million **bytes**. ⇨ **kilobyte**

megatrial   a very large trial. Usually considered to include several thousand subjects

meta-analysis   an analysis of the summary results from two or more similar studies. (Strictly, analyses of analyses; ⇨ **metadata**.) Such methods are becoming more common and are used as a way of synthesising data from a variety of studies to try to get better answers to specific medical questions. ⇔ **overview**

metabolise   to change (when a drug changes in the body). ⇨ **pharmacokinetics**

metabolism   the set of changes that happen to a chemical (a drug) in the body. ⇨ **pharmacokinetics**

metadata   data about data. For example, a manufacturer's data regarding accuracy of a peak flow meter might be considered as metadata

method   a way of carrying out a procedure. The term applies equally to methods of treating patients, methods of measuring variables, methods of analysing data, etc.

methodologist   one who studies and is an expert in **methods**. The term is usually used to distinguish between applied research and theoretical research

methodology   a set of **methods**

me-too   a term used to describe a product for which a very similar alternative already exists

metric   any **measurement scale**, but particularly one referring to **metric data**

metric data   data measured in the SI system of units, which includes grams and metres. Also sometimes used to refer to **continuous data**

metric scale   ≈ **continuous scale**

metric variable $\approx$ **continuous variable**

microcomputer   a computer that is usually small enough to fit on a desk, in a briefcase, or even in a pocket. $\Rightarrow$ **minicomputer, mainframe computer, supercomputer**

microprocessor   the processing unit that forms the basis of a computer

mid *P*-value   an adjustment made to the calculation of *P*-values when working with **ordinal data**. With continuous data, the probability of observing any particular value is considered to be zero; so the probability that x is greater than y [Prob $(x > y)$] is the same as the probability that x is greater than or equal to y [Prob $(x \geq y)$]. However, since with ordinal data any particular value can have a nonzero probability, the mid *P*-value is defined as Prob $(x > y) + \frac{1}{2}$ Prob $(x = y)$

midpoint   the middle of a **class interval**. It is simply the mean of the lower **class limit** and the upper class limit; it is not the median within the class interval

mid-quartile   the mean of the **lower quartile** and **upper quartile**. $\Leftrightarrow$ **median**

mid-range   the mean of the **minimum** value and the **maximum** value

mid-spread $\approx$ **interquartile range**

minicomputer   a small computer that is larger and more powerful than a microcomputer but not as large or powerful as a **mainframe computer** or **supercomputer**

minimax rule   a rule that calculates the maximum value of a **parameter** under different circumstances (often the maximum cost under different circumstances) and then chooses as 'best' the set of circumstances with the minimal cost. It is the minimum of all the possible maxima

minimisation   a **pseudorandom** method of assigning treatments to subjects to try to **balance** the distribution of **covariates** across the treatment groups. $\Rightarrow$ **randomisation, stratified randomisation**

minimum   the smallest of a set of values. $\Leftrightarrow$ **maximum**

minority   less than 50%. $\Leftrightarrow$ **majority**

minus infinity   a number smaller than any other number can be. $\Leftrightarrow$ **infinite, plus infinity**

misclassification   with **categorical data**, misclassification is any form of **measurement error** that ultimately means that a subject is recorded as being in the wrong category. Examples include gross errors such as recording a subject as being male instead of female, or lesser errors such as recording 'partial' improvement of symptoms instead of 'moderate' improvement

misconduct $\approx$ **fraud**

missing at random   missing data where the probability of data missing may depend on the values of some other measured data but does not depend on the missing values themselves. ⇔ **missing completely at random**

missing completely at random   missing data, where the probability of data missing is independent of any observed or unobserved data. This is not very common since subjects may often withdraw from studies because their disease is completely cured or may default because their disease is extremely severe. ⇔ **missing at random**

missing data   a data value that should have been recorded but, for some reason, was not

missing value   ≈ **missing data**

mixed effects model   a statistical model that contains a mixture of different types of **parameters**. Specifically, it is one that contains both **fixed effects** and **random effects**

mixed model   ≈ **mixed effects model**

mock report   ≈ **ghost report**

mock table   ≈ **ghost table**

modal   relating to the **mode**

modal class   in data measured in categories, the most frequently occurring **class**. ⇨ **mode**

modality   the property of having a **mode**

mode   the most frequently occurring value. Used as a measure of **location**. ⇔ **mean, median**

model   an idealistic description of a real (often uncertain) situation. Models may take the form of physical imitations of medical devices, through to mathematical models that are equations or functions describing how a process behaves and on to statistical models that contain both **deterministic** elements (like mathematical models) and **random** elements. Statistical models are often thought of as being like **regression models, logistic regression, log-linear models**, etc. In fact, simple *t* tests are also models, just of a much simpler form. Models can be expressed in words: the model that an independent samples *t* test assumes is that the distribution of a variable is identical in each of two groups, except for a shift in location. Such models can also be expressed algebraically as $y_i = \alpha + \beta x_i + \varepsilon_i$

model equation   the equation for a **model**

modified Fibonacci series   a modification to a standard **Fibonacci series**

moment   a series of **statistics** describing a **probability distribution**. The first moment is the **mean**; the second moment (often referred to as the 'second moment about the mean') is the mean of the squared distances of each value from the mean; the third moment is the mean of the cubed distances of each value from the mean, etc.

monitor   one who visits **investigators** to help with study management, ensure that all data are being recorded as they should be and that all supplies (drugs, materials, etc.) are available on site, and who often returns completed **case record forms** to the data management office. Also a term used for one of the output devices (the screen) of a computer

monitoring committee $\approx$ **data and safety monitoring committee**

monitoring report   a report (usually written) to describe the activities of a **monitor** at a study site and any positive or negative findings, any issues that need bringing to the attention of others, etc.

monotherapy   a single drug. $\Leftrightarrow$ **combination drug**

monotonically decreasing   **repeated measurements** that only remain constant or decrease; they never increase. $\Leftrightarrow$ **monotonically increasing**

monotonically increasing   **repeated measurements** that only remain constant or increase; they never decrease. $\Leftrightarrow$ **monotonically decreasing**

Monte Carlo method   a method to solve a problem by **simulation**

Monte Carlo simulation   either a single **simulation** (as in a **Monte Carlo trial**) or a complete set of simulations forming a **Monte Carlo method**. All Monte Carlo methods are simulations

Monte Carlo trial   one (usually of many thousands) of the simulations in a **Monte Carlo simulation**

morbid   prone to disease. $\Leftrightarrow$ **mortal**

morbid event   an event associated with illness

morbidity   relating to ill health. $\Leftrightarrow$ **mortality**

morbidity curve   a graph of the cumulative occurrence of **morbidity** with time

morbidity rate   the proportion of subjects with a **morbid event** at any given point in time

mortal   prone to death. $\Leftrightarrow$ **morbid**

mortality   relating to death. $\Leftrightarrow$ **morbidity**

mortality curve   a graph of the cumulative occurrence of death with time. $\Leftrightarrow$ **survival curve**

mortality rate   the proportion of subjects who have died at any given point in time

most powerful test $\approx$ **uniformly most powerful test**

moving average   a term used most often with **time series** data. It involves calculating the mean (or average) of observations 1 and 2 (for example); then the mean of observations 2 and 3; then of observations 3 and 4, and so on

multicentre   involving more than one study centre

multicentre research ethics committee   a **research ethics committee** that assesses studies that are planned to take place in many study centres. ⇔ **local research ethics committee**

multicentre study   a study carried out at more than one study centre

multidisciplinary   involving more than one scientific discipline (or speciality). This may include more than one medical discipline (such as oncology and gastroenterology) but also can include other disciplines such as biostatistics (for study design and analysis), mechanical engineering (if prostheses or other **medical devices** are being used), etc.

multidisciplinary study   a study that involves more than one scientific discipline for its design, execution, analysis, and reporting

multilevel model   a model that has a hierarchy to its **parameters**. For example, a study may be conducted in several countries (level 1); with several investigators (level 2) in each country; with many subjects (level 3) recruited by each investigator; and each subject observed on several occasions (level 4). ⇨ **mixed effects model**

multimodal   having more than one **mode**

multimodal distribution   a distribution that has more than one peak (or 'local maxima'). Note that the **mode** is the most frequently occurring value so the term multimodal is a tautology; hence more than one 'peak' is used in this context

multinomial data   ≈ **categorical data**

multiperiod crossover design   a **crossover study** with more than two study periods

multiperiod crossover study   a study designed as a **multiperiod crossover design**

multiple comparison method   any statistical method for making **multiple comparisons**

multiple comparison test   any form of statistical **significance test** for making **multiple comparisons**

multiple comparisons   more than one comparison (usually in the form of statistical **significance tests**) within a single study. The comparisons may be between more than two treatments, or between two treatments but with more than one **response variable**, or a mixture of both of these situations

multiple correlation coefficient (R^2)   the correlation in a **multiple regression model**. ⇨ **correlation coefficient**

multiple dose design   ≈ **repeated dose design**

multiple dose study   ≈ **repeated dose study**

multiple endpoints   more than one **endpoint** in a study. ⇨ **multiple**

comparisons, multiple outcomes

multiple imputation   a method of **imputing** several randomly different values for a **missing value**. The method may have no impact on any **point estimate** over and above that of simple **imputation** but it does better reflect variability of the missing value. ⇨ **last observation carried forward**

multiple linear regression   ≈ **multiple regression**

multiple linear regression model   ≈ **multiple regression model**

multiple logistic regression   **logistic regression** with more than one **covariate**

multiple logistic regression model   a statistical model resulting from **multiple logistic regression**

multiple looks   more than one analysis of **accumulating data**. ⇨ **group sequential study**

multiple outcomes   a study having more than one **outcome variable**. ⇨ **multiple endpoints, multiple comparisons**

multiple regression   **linear regression** with more than one **covariate**

multiple regression model   a statistical **regression model** resulting from **multiple regression**

multiple significance tests   the use of more than one statistical **significance test** in one study

multiplicative model   a statistical model where the combined effect of separate variables contribute as the product of each of their separate effects. ⇔ **additive model, linear model**. ⇨ **interaction**

multiplicity   ≈ **multiple comparisons**

multistage design   a study that has more than one stage (or period), possibly including a **run in period**, a **treatment period** and a **follow-up period**

multi-univariate   more than one **univariate** response variable where the interest lies with each variable in its own right, rather than a **multivariate** combination of them

multivariate   relating to more than one variable (usually more than one **response variable**). ⇔ **univariate**. ⇨ **bivariate**

multivariate analysis   special methods of analysis suitable for **multivariate data**

multivariate data   measurements that consist of more than one variable. For example, a person's 'size' could be measured jointly by their femur length, tibia length and skull circumference. More than two variables are always referred to as multivariate: two variables, whilst still multivariate, are often referred to as **bivariate**

multivariate distribution   the **joint distribution** of more than one variable. ⇔ **univariate distribution**. ⇨ **bivariate distribution**

Münch's law   a pessimistic rule which suggests that the number of subjects expected to be available to enter a study should usually be divided by a factor of at least ten. ⇨ **Lasagna's law**

mutually exclusive   not able to occur at the same time

mutually exclusive events   two or more events that are not able to occur at the same time. This is not restricted to situations where events have (by chance) not been observed to occur at the same time, but events that are not capable of jointly occurring. An example would be that a subject is male and is pregnant

named patient use   a way of allowing a doctor to offer an unlicensed product to a patient outside of a clinical trial. This is often allowed in treatment of life threatening diseases where no alternative treatment is available. There are strict guidelines under which such a supply may be offered. ⇨ **compassionate use**

natural experiment   a term used to describe a (usually major) event (usually some form of disaster). The resulting change in environment and its impact can be studied. It is not a true **experiment** as the intervention is not under our control. Examples include floods and chemical leaks

natural history   the course of events over time. The term can be used on a massive scale to describe geological and climatic changes or to describe how an illness in an individual has developed and is likely to develop over time. ⇔ **medical history**

natural logarithm   a logarithm to base e. ≈ $\log_e$

necessary and sufficient   a term often used in a mathematical context but applicable elsewhere. It describes a set of circumstances that are required ('necessary') but also where no other circumstances are simultaneously required ('sufficient'). For life to exist, oxygen must be present; but that is not all. Thus, oxygen is necessary, but not sufficient, for life to exist

negative control   see **inactive control, placebo control**

negative correlation   **correlation** between two variables such that as one variable increases the other tends to decrease. ⇔ **positive correlation**. ⇨ **inverse correlation**

negative effect   an effect that is undesirable. ⇨ **adverse reaction**

negative gain   a loss. ⇔ **negative loss**

negative loss   a gain. The term is used when referring to several items that generally incur a loss (possibly a financial loss). To avoid switching between losses and gains, the term negative loss is sometimes used. ⇔ **negative gain**

negative predictive value   in a **diagnostic test**, the probability that a person with a negative result does not have the disease (a correct result). ⇨ **positive predictive value, sensitivity, specificity**

negative relationship   an informal term for **negative correlation**. ⇔ **positive relationship**

negative response   a poor response, or no response, to treatment. ⇔ **positive response**

negative result   a result less than zero. The result of a **negative study**. ⇔ **positive result**

negative skew   describes a **distribution** that has a long left hand **tail** so that the majority of observations are at the upper end of the scale. ⇔ **positive skew**

negative study   a study that fails to reject the **null hypothesis** or otherwise fails to fulfil its objectives. ⇔ **positive study**

negative treatment effect   an undesirable treatment effect. ⇨ **adverse reaction**. ⇔ **positive treatment effect**

nested design   an experimental design where some **factors** occur only as subsets of other factors. ⇨ **multilevel model**

nested factor   a **factor** that occurs in an experiment only as a subset of another factor

net change   any change after removing the effect in a control group. For example, if we calculate the change in blood pressure in a group of treated patients and in another group of untreated patients, the net change in the treated group would be their change minus any change observed in the untreated group. ⇨ **treatment effect**

net difference   ≈ **net change**

net effect   any effect after removing some baseline or control effect. ⇨ **net change**

net treatment effect   see **net change, treatment effect**

network   a system of communication between computers or people to share information

new chemical entity (NCE)   a new chemical that is being developed as a potential new drug

new drug application (NDA)   an application to the Food and Drug Administration in the USA for a licence to market a **new chemical entity**

Newman–Keuls test   a **multiple comparison test** for testing the **null hypothesis** of no difference between the means of more than two groups

next of kin   a person's nearest relation (through either blood or marriage). ⇔ **legal guardian**

nil effect   no effect, or zero effect

no cause audit   an **audit** carried out as a matter of routine or because a study or a site has been selected at random for audit. ⇔ **for cause audit**

node   on a **decision tree** (Figure 6), any point at which a choice of routes can be made

***n*-of-1 study**   a study carried out in a single patient to determine the best treatment for that patient (which may not necessarily be the best treatment for patients in general)

noise   unwanted variation in data. ⇨ **signal to noise ratio**

noisy data   data that have a lot of **noise**, or a high **variance**

nomenclature   the terminology (symbols and special language) used in any science or discipline

nominal data   ≈ **categorical data**

nominal scale   a **categorical scale** whose possible values are simply in the forms of names: country of origin, concomitant medications, etc.

nominal variable   a variable measured on a **nominal scale**

nomogram   a type of graph used to depict the relationship between (usually) three variables (Figure 21)

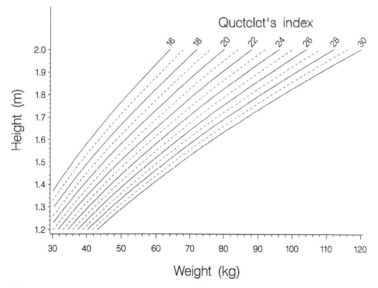

Figure 21   Nomogram. For values of height and weight, Quetelet's index (body mass index) can be read off

noncentral distribution   a variation of the more standard **probability distributions** (**t distribution**, **F distribution**, etc.) useful for calculating **power** of **significance tests**

noncompliance   the act of not fully complying with a protocol. Often the term is restricted to whether or not a subject takes the medication as and when they should but it can be interpreted more widely to any aspect of a protocol

noncompliant   a subject who does not fully comply with a protocol

nonignorable missing data   **missing data** that indicate something about the subject because of the fact that the data are missing. For example, in an antihypertension study, data may be missing because a patient died: a death caused by a road traffic accident may be considered ignorable because it is unlikely to be study related but a death caused by a cardiac arrest would not be ignorable. In this example the term is partially related to a contrast between **adverse events** and **adverse reactions**

nonignorable missingness   a process that produces **nonignorable missing data**

noninferiority study   a study whose objective is to show that one treatment 'is not worse than another'. This is subtly different to showing that two treatments are equivalent ($\approx$ **equivalence study**) and obviously different to trying to show that one treatment is different to another ($\approx$ **difference study**). $\Rightarrow$ **superiority study**

noninformative censoring   **censoring** in **survival studies** that is completely unrelated to treatment. Essentially the same meaning as **nonignorable missing data** in the context of survival studies and censoring

noninformative missing data   data that are missing, and the fact that they are missing tells us nothing about what the data value should be. $\Rightarrow$ **missing completely at random**

noninformative prior $\approx$ **reference prior**

noninvasive   any medical procedure that is not **invasive**

nonlinear   not in a straight line. $\Leftrightarrow$ **linear**

nonlinear model   a **model** that contains **multiplicative** terms, not simply **additive** terms. $\Leftrightarrow$ **linear model**

nonparametric   a branch of statistics that makes few **assumptions** about the **distributions** of data

nonparametric data   strictly, there is no such thing as nonparametric data. However, the term is quite commonly used to refer to data that come from **distributions** that do not obviously resemble any standard **probability distribution** and for which **nonparametric methods** of analysis need to be used. $\Leftrightarrow$ **parametric data**

nonparametric method   any statistical method for **significance testing** and estimation that makes fewer assumptions about the distribution of the data than do **parametric methods**. It is widely believed that these methods make no assumptions at all about the distribution of the data but this is not the case

nonparametric test   a **nonparametric** statistical **significance test**. Examples include the **Mann–Whitney** *U* **test**, the **Wilcoxon matched pairs signed rank test**, etc.

nonrandom   not **random**; used to refer to nonrandom **samples** and nonrandom **treatment allocation**

nonrespondent   a subject who does not answer a question, either because they refuse to or because they did not attend a study visit and so could not be asked

nonresponse   similar meaning to **nonrespondent** but also used to describe subjects who do not respond to treatment

nonsense correlation   an observed **correlation** that may be **statistically significant** but which does not make any biological or medical sense in terms of **causality**

nonsignificant risk study   a study of a **medical device** that poses no important risk to the subjects who take part. ⇔ **significant risk study**

nonzero effect   this is usually used to refer to an effect when it needs to be stressed that an effect *does* exist. This may be because the effect is very large or because, despite the effect being very small, it may still be medically or scientifically important

normal   a rather dangerous term: it has an everyday use meaning typical or not unusual; it has a similar meaning in a technical sense of a **normal range** (⇨ **reference range**) for a variable; it also has a highly technical (statistical) use as in **Normal distribution**, one of the most basic ideas in statistics. Because of these diverse uses, it is important to either avoid its use altogether or to be highly specific. In this book, an upper case 'N' is used for the statistical **probability distribution**, the Normal distribution

normal approximation   an approximate procedure based on assuming data come from a **Normal distribution**

normal curve   an informal term used to describe the shape of the curve of a **Normal distribution**

Normal distribution   the **probability distribution** that is very commonly used (either directly or as a basis for further refinements) in statistical **significance testing**, **estimation**, model building, etc. (Figure 22)

normal limit   the upper (or lower) limit of a **normal range**

normal plot   ≈ **quantile–quantile plot**

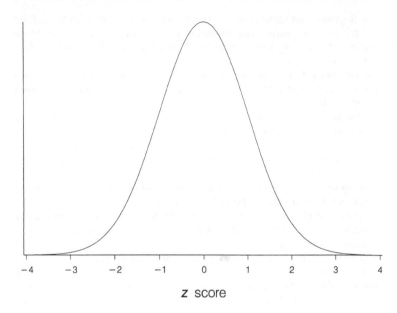

z score

Figure 22   The classic 'bell shape' of a Normal distribution with mean 1 and standard deviation 1

normal range   the usual range within which the values of a variable can be expected to lie. It usually implies that all subjects within that range will be healthy. ⇔ **reference range**

normality   the degree to which a **distribution** is like a **Normal distribution**

normally distributed   said of a set of data that come from an underlying **Normal distribution**

not significant   either an effect that is of no clinical importance (≈ **clinically significant**) or one that, regardless of its size, is not **statistically significant**

notifiable disease   a **disease** that must, by law, be notified to health authorities

nuisance parameter   in a statistical model, **parameters** that may be very important as **covariates** but which are not of direct interest in the study. Usually the **treatment effect** is of most interest; if it turns out that subject's age or previous history are predictive of **outcome** (but equally predictive within each treatment group) then their parameters would be considered as nuisance parameters

nuisance variable   any variable in a statistical model that is not of primary interest. ⇨ **nuisance parameter**

null distribution   the **probability distribution** of a variable if the **null hypothesis** is true

null hypothesis (H_0)   the assumption, generally made in statistical **significance testing**, that there is no difference between groups (in whatever parameter is being compared). Evidence (in the form of data) is then sought to refute (or reject) this null hypothesis. ⇔ **alternative hypothesis (H_1)**

number needed to harm   the number of patients that a physician would have to treat with a new treatment in order to harm (in some predefined sense) one extra subject who would otherwise not have been harmed. 'Harm' may be in the context of a treatment failure, an **adverse reaction**, a death, etc. More usually considered in the context of **number needed to treat**

number needed to treat   the number of patients that a physician would have to treat with a new treatment in order to avoid one **event** that would otherwise have occurred with a **standard treatment**

numerator   in a fraction, such as $\frac{1}{2}$ or $\frac{3}{4}$, the numerator is the number on the top line of the fraction (in these cases 1 and 3, respectively). ⇔ **denominator**

numeric   relating to numbers only. ⇔ **alphanumeric**

numeric variable   a variable that is a number. This generally means it is a **continuous variable** and not, for example, a **likert scale**

Nuremberg Code   a set of **ethical** principles about research on humans that formed the basis of the **Declaration of Helsinki**

n-way   a generalisation of 1–way, 2–way, 3–way, etc. meaning any number of ways. Used particularly in the sense of **n-way analysis of variance**, **n-way classification**, etc.

n-way analysis of variance   a generalisation of **analysis of variance** indicating that many (n) **factors** are included in the model

n-way classification   classification of a **continuous variable** (usually by a **discrete variable**) in several (n) **subclasses**

O'Brien and Flemming rule one of the most common **stopping rules** used in **group sequential studies**. ⇨ **Pocock rule**

objective the purpose of a study. It may be described either in very precise and specific terms or in general terms such as 'to assess the safety and efficacy of Drug A'. The term is also used to refer to clear facts rather than general impressions. For this interpretation ⇔ **subjective**

objective data data that are usually considered to be measured with high **accuracy** and that have low (or negligible) **intraobserver variation** and **interobserver variation**. ⇔ **subjective data**

objective endpoint an endpoint to a study that is **objective data**. ⇔ **subjective endpoint**

objective measurement a measurement of **objective data**. ⇔ **subjective measurement**

objective outcome an **outcome** that is **objective data**. ⇔ **subjective outcome**

observation usually meaning the data relating to one of the subjects being studied. However, the term is mostly used in a computing context to mean the number of rows in a (rectangular) **database**. Usually this will consist of one observation (with many variables) per subject; sometimes, if there is a different number of records per subject, the database may be set out as one observation per record

observational study a study that has no experimental intervention but just observes what happens to a group of subjects. ⇨ **case-control study**, **cohort study**. ⇔ **intervention study**

observed change the change in a variable that is seen to occur. This is in contrast to a **fitted value** from a statistical model

observed data strictly this is synonymous with **data** but use of the term 'observed' helps to contrast with **fitted values** from statistical models

observed difference the difference (in means, proportions, etc.) for a variable that is seen to occur. This is in contrast to a **fitted value** from a statistical model

observed distribution ≈ **observed frequency distribution**

observed effect   usually the simple estimate of an **effect** (difference in **means**, difference in **proportion**, the **odds ratio**, etc.) that has not been **adjusted** to account for any possible **covariates**

observed frequency   the **frequency** with which a specific variable is seen to occur. This is in contrast to a **fitted value** from a statistical model

observed frequency distribution   strictly this is synonymous with **frequency distribution** but use of the term 'observed' helps to contrast with **probability distribution**

observed mean   the **sample mean**. ⇔ **population mean**

observed outcome   the observed value (usually of **categorical data**). This is in contrast to an **expected outcome** from a statistical model

observed rate   the **rate** at which an **event** is seen to occur. This is in contrast to any **fitted values** from statistical models

observed relative frequency distribution   the **observed frequency distribution** presented as a **relative frequency distribution**

observed result   any kind of result that is seen to occur. This is in contrast to any **fitted values** from statistical models

observed sample size   the **sample size** actually obtained, in contrast to what was planned

observed treatment difference   ≈ **observed effect**

observed treatment effect   ≈ **observed effect**

observed value   either the value of a measurement in a single subject or the number of occurrences of an event that have been observed

observed variance   the **sample variance**. ⇔ **population variance**

observer bias   any **bias** in measurements introduced by an observer (for example **digit preference**) or caused by making observations (≈ **Hawthorne effect**)

observer error   any error in measurements made by an observer. ⇨ **intraobserver agreement, interobserver agreement**

observer variation   see **interobserver variation, intraobserver variation**

Occam's razor   a philosophical stance which prefers simple explanations to more complex alternatives. This is a general principal to adopt in formulating statistical models

Ockham's razor   ≈ **Occam's razor**

odds   the **probability** of an event occurring divided by the probability of it not occurring. For example, if one in ten cancer patients are cured by a drug, then the odds of being cured are stated as 1:9. ⇔ **rate, risk**

odds ratio   the ratio of two **odds**, often used as a summary of the size of a treatment effect in **two-by-two tables**. In Table 12, the odds ratio is calculated as $(37 \times 31) \div (13 \times 19) = 4.6$. ⇔ **risk ratio**

Table 12   Contingency table showing the distribution of treatment response by treatment group

	Treatment A	Treatment B	
Treatment success	37	19	
Treatment failure	13	31	
Total	50	50	

off label   the use of a product to treat a disease for which it does not have a **marketing authorisation**

off site   away from the buildings or facilities where key activities occur. This may be with reference to study medication being stored at a location separate from where patients are treated or it may refer to an **archive** of data being kept at a location separate from where the main data-processing activities take place. ⇔ **on site**

off study   refers to clinical activities that may occur concurrently with a study protocol but which are not included in the **protocol**, or to procedures which take place, or medication that is given, after a subject has completed the protocol. ⇔ **on study**

off treatment   any time (during the course of a study or after a subject has completed a study) when a subject is not being given treatment (or placebo). This may be during a **run in period** or during a **long term follow-up** period. ⇔ **on treatment**

ogive   a graph of a **cumulative frequency distribution** (Figure 23)

ointment   a **vehicle** for delivering **topical** treatment, usually paraffin or Vaseline based. ⇨ **cream, gel, lotion**

omitted covariate   a **covariate** that has not been included in a **regression model** or **analysis of covariance** model (either intentionally or inadvertently)

omnibus test   any statistical **significance test** that involves comparing **parameters** (often means or proportions) from more than two groups. It may, for example, be a test that all the means are equal: in such a case, if the **null hypothesis** (of equal means) is rejected we cannot immediately say which means are different to which others

on site   activities that take place (or the availability of study material) at the site where they are needed. This may relate to medication being on the site where patients are treated, or to completed **case record forms** being at the premises of the data management office. ⇔ **off site**

on study   activities that take place as part of a study **protocol**. ⇔ **off study**

on treatment   any time when a subject is being given a study treatment (or placebo). ⇔ **off treatment**

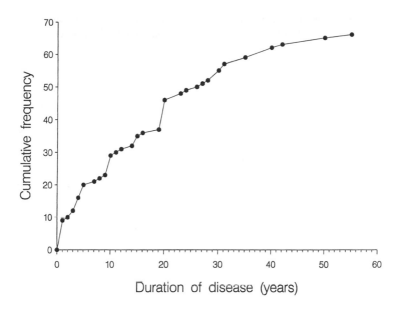

Figure 23   Ogive. A graph of the cumulative number of patients who have suffered from eczema for less than 1 year (9 patients), less than 2 years (10 patients), less than 3 years (12 patients), . . . less than 55 years (all 66 patients)

one sided   concerned with only one **tail** of a **distribution.** ⇔ **two sided**

one sided alternative   the **alternative hypothesis** that is a **one sided hypothesis.** ⇔ **two sided alternative**

one sided hypothesis   a **hypothesis** that allows for the possibility of a difference in only one direction (for example, Drug A must be better than Drug B). Such hypotheses are not as common as **two sided hypotheses**

one sided test   any statistical **significance test** that will accept a **one sided hypothesis** if the **null hypothesis** is rejected. ⇔ **two sided test**

one tailed   ≈ **one sided**

one tailed alternative   ≈ **one sided alternative**

one tailed hypothesis   ≈ **one sided hypothesis**

one tailed test   ≈ **one sided test**

one way analysis of variance   the simplest form of **analysis of variance,**

used to compare the means of two (or more) groups in a **parallel groups study** but without including any other **factors** or **covariates** in the statistical model

one way classification   data that are grouped by only one **categorical variable**. Note that the categorical variable may have several levels ($\approx$ **levels of a factor**) but there is only one variable

one way design   a study design that involves only a **response variable** and one (**categorical**) **covariate**

online   a computing term meaning that work is being done directly onto a central computer rather than being temporarily held on a local computer before **batch processing** or being **uploaded** to the central computer

online data entry   electronic **data entry** that occurs **online**. $\Leftrightarrow$ **distributed data entry**. $\Rightarrow$ **remote data entry**

open class interval   a **class interval** that either has no lower limit (it is all values below a certain value) or has no upper limit (it is all values above a certain value). It is often used with highly **skewed data**

open label   not **blind**

open label study   a study where the treatments are not **blinded**

open sequential design   a **sequential study** design that does not have any upper limit to the number of subjects that may be recruited (Figure 24). $\Leftrightarrow$ **closed sequential design**

open sequential study   a study that is designed as an **open sequential design**

open study $\approx$ **open label study**

open treatment assignment   treatment assignment that is not **blinded** (although it may still be random). $\Rightarrow$ **open label study**

operation   a surgical procedure or a mathematical function

optimal design   a study that is the best ('optimal') for some specific purpose. Note that it may not be optimal for all purposes. It may be optimal on statistical grounds or from practical study management grounds

oral assent   assent that is given orally. $\Leftrightarrow$ **written assent**. $\Rightarrow$ **consent**

oral consent   consent that is given orally. $\Leftrightarrow$ **written consent** (which is more common). $\Rightarrow$ **assent**

order of magnitude   a multiple of, or division by, 10

order statistic   any one of the **centiles**

ordered   see **ascending order**, **descending order**

ordered alternative hypothesis   an **alternative hypothesis** that involves more than two groups. The simplest example is that of comparing the means of three groups. The **null hypothesis** is that 'all the means are

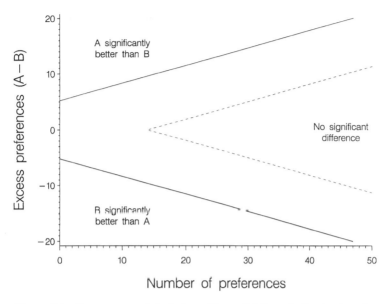

Figure 24 Open sequential design. The solid lines indicate stopping boundaries for declaring a statistically significant difference between treatments A and B. If the broken boundary is crossed, then the study stops, concluding that no significant difference was found between the treatments. Potentially, the number of preferences could continue indefinitely between the upper solid and broken lines or between the lower solid and broken lines; in such a case no conclusion would ever be reached

equal' or, equivalently, '$\mu_A = \mu_B = \mu_C$'; the simplest alternative hypothesis might be that 'not all of the means are equal'; an ordered alternative hypothesis would be that '$\mu_A < \mu_B < \mu_C$'

ordered categorical data   data that are measured on a **categorical scale** but where the categories have a natural ordering, for example mild, moderate and severe. ⇨ **likert scale**

ordered categorical scale   the scale on which **ordered categorical data** are measured

ordered categorical variable   a variable that yields **ordered categorical data**

ordered data   data that are measured on an **ordered scale**

ordered logistic regression   an extension of the methods of **logistic**

regression where the **response variable** is **ordered categorical,** instead of binary. ⇨ **polytomous regression**

ordered scale   a **measurement scale** that is **ordered.** This includes **ordinal scales, ordered categorical scales, interval scales**

ordinal data   data that are simply **ordinal numbers**

ordinal number   the numerical position (1st, 2nd, 3rd, etc.) in a set of **ordered data**

ordinal scale   the scale on which **ordinal data** are measured

ordinal variable   a variable that yields **ordinal data**

ordinary least squares   ≈ **least squares**

ordinate   ≈ y **axis.** ⇔ **abscissa** (or x **axis**)

orientation   layout, generally of paper in the form of either **landscape** or **portrait**

origin   the point of zero on a graph. On a two-dimensional graph, where $x = 0$ and $y = 0$

original data   ≈ **source data**

original document   the top copy (not photocopies, etc.) of a document. ⇨ **source data**

original record   ≈ **source data**

orphan drug   a product that has a limited market because it is used for a rare disease. Regulatory requirements are different for orphan drugs than for non-orphan drugs

orthogonal   when two ideas, measurements, estimates, etc. are at right angles to each other, implying that they are also **independent** of each other

orthogonal contrasts   two (or more) **contrasts** that are **independent** of each other

outcome   usually the **primary variable** of a study. Although an outcome would generally be an event (≈ **outcome event**), the term is frequently used to refer to the primary variable whatever the **measurement scale**

outcome event   the **primary variable** of a study, specifically when that variable is **binary**

outcome measure   the **primary variable** of a study, usually restricted to the case when that variable is **continuous**

outcome variable   the variable that defines the **outcome** for a study

outcomes research   methods of trying to answer research questions that do not involve **intervention studies** but which analyse **databases** and attempt to control for all possible **confounding factors** by complex statistical modelling. It is a cheaper, quicker and relatively easier way to answer a question than doing a clinical trial and so presents clear advantages over trials (which are often expensive and time consuming).

However, much of the rigour and control of **bias** gained from clinical trials may be lost

outlier   a data value that does not seem to be true, given all the other data values, usually because it is very **extreme** (either too large or too small). ⇔ **inlier**

outpatient   a patient who is not kept in hospital overnight. Note that treatment may still be given in the hospital. ⇔ **inpatient**

outpatient study   a study of **outpatients**

output device   a method of getting data out of a computer (this may simply be the monitor or a printer)

over represent   when there is a higher proportion of some **subgroup** in a sample than there is in the population. Sometimes this may be desirable. The proportion of subjects with mild, moderate and severe symptoms may intentionally be kept equal in a sample, even though they are not similar in the population. ⇔ **under represent**. ⇨ **sample demographic fraction**

overmatch   in **matched studies**, **cases** and **controls** might be matched for as many variables as is reasonably possible. If, however, the **exposure variable** is also matched between the groups then no difference between the groups will be found. This is called overmatching and is a risk in complex **epidemiological studies** where the variable causing the cases is not known

over-the-counter drug   products that can be purchased without needing a doctor's **prescription**. ⇔ **prescription only medicine**

overview   to look at data from various sources, considering them as a whole and making a conclusion. ⇔ **meta-analysis** that involves more formal assessments of the completeness of the data and more formal statistical methods for combining them. Overviews and meta-analysis are very important methods for synthesising data

package see **computer package, package insert**

package insert the information given to a patient with a pack of medication. It contains information similar to the **summary of product characteristics** but is written in a style appropriate for patients to understand

page orientation ≈ **orientation**

pair two items. Usually this means the same variable measured on two similar subjects or the same variable measured on one subject on two occasions. It can sometimes refer to two independent items that are brought together in some way (≈ **pairwise comparisons** for an example)

pair matching ≈ **pairwise matching**

paired comparison a **comparison** that is made on **paired data** (not on **independent groups**). ⇔ **pairwise comparisons**

paired data the same variable measured on two similar subjects or the same variable measured on one subject on two occasions

paired design a study design that involves taking **paired observations** and usually makes treatment comparisons using **paired comparisons**, often (but not necessarily) in the form of a crossover design

paired observations two observations that are related to each other, either as two observations from the same subject at different times (or on different sites on the body) or as one observation from each of two matched subjects in a **paired design**

paired sample a sample of **paired observations**

paired *t* test a statistical **significance test** testing the **null hypothesis** that the mean difference in a population (from which a sample of **paired data** has been taken) is equal to some particular value. Usually it is to compare the mean difference with zero. ⇔ **independent samples *t* test**

pairwise relating to **pairs**

pairwise comparisons in a study where more than two groups are being compared, the term pairwise refers to each of the possible **pairs** of treatments that can be compared. For example, when there are three groups (A, B, C) there are three possible pairwise comparisons: A *vs.* B, A *vs.* C, and B *vs.* C. ⇨ **multiple comparison method**

pairwise matching   matching a **pair** of subjects

palliative care   care for the whole patient, rather than treatment of specific symptoms. Examples include supportive care in the form of good communication, sympathy, understanding, empathy, etc. towards patients (and their relatives)

pandemic   occurring over a large geographic area. ⇔ **endemic**

paperless   using no paper; as in 'paperless case record form' (where data are entered directly onto a computer without being transcribed onto paper first)

parallel   side by side; not crossing over

parallel assay   a **dose finding study** where the activity of a new product is compared with the activity of a standard drug

parallel control   the **control group** in a **parallel group study**

parallel dose design   a **parallel group study** where the different groups of subjects receive different doses of the same drug

parallel group design   the most common design for clinical trials, whereby subjects are allocated to receive one of several treatments (or **treatment regimens**). All subjects are independently allocated to one of the **treatment groups**. No subjects receive more than one of the treatments. ⇔ **crossover design**

parallel group study   a study designed as a **parallel group design**

parallel study   ≈ **parallel group study**

parallel track   occurring at the same time but independently. For example, two studies that are being conducted at the same time but independently of each other

parameter   the true (but often unknown) value of some characteristic of a **population**. A simple example is the mean age of a population. Other examples include variances, minimum values and medians. Parameters are usually denoted by Greek letters (for example, σ^2 for the variance) and are estimated by **sample statistics** that are usually denoted by Roman letters (for example, s^2 for the variance). The most common parameter that we wish to estimate in clinical trials is the size of the **treatment effect**

parameter estimate   the estimate (based on data) of a **parameter**

parametric data   as with **nonparametric data**, this term has no real meaning but it is quite commonly used to refer to data that come from recognisable **probability distribution** and for which **parametric methods** of analysis can be used

parametric method   statistical methods that make specific assumptions about the **distributions** of data. Examples include the *t* **test**, **correlation** and **regression**. ⇔ **nonparametric method**

parametric test   any statistical **significance test** that uses parametric methods. Examples include the *t* **test** and the *F* **test**

parent   this term is used in the obvious way referring to mothers and fathers of children. It is also sometimes used in **decision trees** and mathematical models. In decision trees, it refers to a **node** from which **branches** come; in mathematical models it sometimes refers to a broad set of models from which other, simpler, models can be formulated. These are the most common uses but the term is sometimes used in any context where a hierarchy exists

parent drug   the basic form of a drug from which various alternative modifications are available

parsimony   the concept of simplicity being preferred over complexity. With particular reference to statistical models, models with few parameters are generally preferred over those with many parameters. ⇨ **Occam's razor**

partial response   in cancer studies, this is generally regarded as a decrease in tumour size of at least 50%. ⇨ **complete response**, **stable disease**, **progression**

partially balanced block   a **block** of treatments that is **balanced** for some **comparisons** but not for others. For example, a block containing two assignments to Treatment A, two to Treatment B and three to Treatment C is only partially balanced

partially balanced design   a study design that uses **partially balanced blocks** of treatments

partially confounded   the situation where two **estimates** are not completely **confounded** but where some information in one estimate is not independent of another. This is very common in unbalanced designs and when using **analysis of covariance**

participant   someone who takes part (usually in a study)

partition   to split up. The term is most usually used when trying to decide if relationships (for example, dose–response relationships) are **linear** or **quadratic**. In this instance, we often refer to partitioning the **sums of squares**

patent   the process of registering, or the documents confirming, ownership of an invention (such as a new drug), thus protecting that invention from being copied

pathogen   a microbiological organism that is capable of causing disease

pathogenesis   the cause and subsequent development of a disease

pathology   the science of the causes of disease

patient   a subject who has a disease or other illness. Note that the requirement to have a disease or other illness differentiates from the broader term 'subject'. Note also that 'volunteer' is not a good choice of word when describing those who take part in studies because healthy subjects and diseased patients should all be taking part voluntarily

patient accrual   ≈ **patient enrolment**

patient chart   any kind of chart or graph on which a patient's data are plotted

patient compliance   the degree to which an individual patient complies with the study protocol as a whole, or specifically complies with taking the appropriate medication

patient contact   any type of meeting between a patient and a health worker. The contact may be face to face, by telephone, by letter, etc.

patient enrolment   the process of recruiting patients into a study

patient enrolment period   the time period during which patients are enrolled into a study

patient follow-up   the process of observing a patient over time, after they have been given study medication. ⇨ **follow-up data, follow-up period, follow-up visit**

patient home visit   a visit (usually by a study nurse or an investigator) to a patient, in the patient's home

patient id   ≈ **subject id**

patient identification number   ≈ **subject identification number**

patient information booklet   a small booklet given to subjects, before they agree to take part in a study, to give them information about the study to help them decide if they are prepared to volunteer

patient information sheet   a smaller form of a **patient information booklet** that is just a single sheet of paper

patient monitoring   observation of a patient to ensure safety (primarily) and sometimes to record efficacy data

patient population   the entire (theoretical) **population** of patients that could be recruited into a study. The term is also used to refer to the different **analysis populations (intention-to-treat population, per protocol population, safety population,** etc.)

patient record   the data referring to a single patient

patient recruitment   ≈ **patient enrolment**

pay journal   a **journal** for which the cost of publication has to be met (fully or partially) by the authors of the manuscripts. **Peer review** may, or may not, also be required. ⇔ **peer review journal**

peak   any area on a graph that shows a rise and subsequent fall

peak value   the maximum value from a set of related data. Usually it is from data that are all from one subject

Pearson chi-squared statistic   ≈ **chi-squared statistic**. The term is often prefixed with 'Pearson' to distinguish it from other forms of statistical **significance tests** that also use the **chi-squared distribution**

Pearson product-moment correlation coefficient   ≈ **correlation coefficient**

Pearson residual   **residuals** in **contingency tables** (and in **logistic regression** models). Each residual is calculated as the difference between the **observed value** and **expected value**, divided by the square root of the expected value

peer   a colleague or other person who is considered an equal in scientific merit and experience

peer review   when an independent scientist of similar standing and experience to the first reviews a manuscript or other documents or working practices and makes comments. ⇨ **expert review**

peer review journal   a **journal** that sends submitted manuscripts for **peer review**. It is usually assumed that there is no charge for publishing an accepted manuscript. ⇔ **pay journal**

per protocol analysis   the analysis of study data that excludes data from subjects who did not adequately comply with the study protocol. ⇔ **intention-to-treat**

per protocol population   the subset of subjects recruited into a study who are included in the **per protocol** analysis

per unit cost   the extra cost incurred (per person treated, per bottle manufactured, etc.) It does not include basic set up costs. ⇔ **fixed cost**

percent   of 100. For example, the phrase '37 percent of patients responded to treatment' means that 'of every 100 patients given treatment, 37 responded'

percent difference index   the difference between two percentages. ⇨ **percentage point**

percentage point   the term is often used in a similar way to **percent difference index**: when considering a change from, for example, 20% to 30%, this can be described as a '50% increase' or a 'difference of 10 percentage points'

percentile   ≈ **centile**

percentile–percentile plot   ≈ **quantile–quantile plot**

per comparison error rate   ≈ **comparisonwise error rate**

per experiment error rate   ≈ **experimentwise error rate**

performance measure   any measurement of how well a person or group of people carried out a particular task; or a measurement of how well an experiment or measuring instrument does what it is intended to do

performance monitoring   the process of reviewing performance (that is, how well a task is being done), with a view to making improvements, if necessary. The term can equally well apply to reviewing the performance of people or machines

period   an interval of time. In the specific context of **crossover studies**, the term refers to the intervals of time when a subject is given the first treatment (period 1), when they are given the second treatment (period 2), etc.

period effect   any **systematic difference** in response between two **periods**. Most commonly used in the context of **crossover studies**

period prevalence   the **prevalence** (number of **cases**) of an event during a specified period of time. ⇔ **point prevalence**

periodic safety update report   a regular report sent to a regulatory authority with details of all **adverse events** reported for a product

peripheral   of secondary importance. In computer terms, it refers to any additional piece of hardware that can be added to a computer (image scanners, printers, etc.)

permutation   any ordering of a given fixed set of data values

permutation test   ≈ **nonparametric test**

permute   to rearrange a set of data values to form a new **permutation**. ⇨ **randomise**

permuted block   ≈ **randomised block**

personal computer   typically a small (although possibly quite powerful) computer. Such computers are sufficiently small that they easily fit onto a desk; some are small enough to fit into a small briefcase. ⇔ **mainframe computer**

personal data   data about individual people. Often it is restricted to data that may be considered as of a sensitive nature (sexual behaviour, illegal substance abuse, etc.)

personal probability   in **Bayesian** statistics, this is one person's **prior probability** of an event. It is sometimes called a 'personal probability' to emphasise that different people may legitimately have different prior probabilities for the same event, so the prior probability is of a 'personal' nature

person-time   see **person-year** as an example. 'Time' can be any chosen units

person-year   when many people have been exposed to an intervention for varying lengths of time, the total time of exposure for all people can be calculated and expressed as if it were one individual exposed for this total length of time. For example, two people each exposed for 6 months would equate to one person-year; one person exposed for 12 months and another with zero exposure would also equate to a total exposure of one person-year

pessary   a **suppository** inserted into the vagina

pharmaceutical   relating to drugs. ⇔ **biologic, phytomedicine**

pharmaceutical company   a commercial organisation that researches, develops, manufactures and markets drugs

pharmaceutical industry   pharmaceutical companies and other support companies involved in the research, development, manufacture and marketing of drugs

pharmacist   a person qualified to prepare, safely store and dispense drugs

pharmacodynamics   broadly, the action of a drug on the **physiology** of the body. ⇔ **pharmacokinetics**

pharmacoeconomics   the study of economic implications of drug usage. This can be used either to try to justify use of drugs as an economic benefit or to evaluate the cost associated with a patient having a disease compared with the cost needed to treat the patient

pharmacoepidemiology   the study of drug usage and results (positive and negative) in broad **populations** with a view to a better understanding of beneficial drug usage. ⇨ **epidemiology, outcomes research, pharmacovigilance**

pharmacogenetics   the study of how drugs affect the genetic makeup of the body

pharmacokinetics   broadly, the action of the body on a drug. Pharmacokinetics includes the study of the rate of **absorption** and **distribution** of products into and around the blood stream, and the rate (and methods) of **elimination** of drug from the body. ⇔ **pharmacodynamics**

pharmacology   the study of drugs (including uses, benefits, harmful effects and stability)

pharmacovigilance   the study of **adverse events** (presumed to be related to drug usage) in broad populations

pharmacy   a place where drugs are stored in secure conditions and under the control of a **pharmacist**

phase   different stages of drug development and testing (≈ **Phase I, II, III, IV study**) or, used on its own, to denote different stages within a study. In this latter case ≈ **phase of study**

Phase I study   the earliest types of studies that are carried out in humans. They are typically done using small numbers (often less than 20) of healthy subjects and are to investigate **pharmacodynamics, phar-**

macokinetics and toxicity

Phase II study   studies carried out in patients, usually to find the best dose of drug and to investigate safety. This term is sometime split into the subgroups **Phase IIa studies** and **Phase IIb studies**

Phase IIa study   of a set of **Phase II studies**, the earlier ones (often on fewer patients) are sometimes referred to as Phase IIa. ⇔ **Phase IIb study**

Phase IIb study   of a set of **Phase II studies**, the later ones are sometimes referred to as Phase IIb. ⇔ **Phase IIa study**

Phase III study   generally these are major studies aimed at conclusively demonstrating efficacy. They are sometimes called **confirmatory studies** and (in the context of **pharmaceutical companies**) typically are the studies on which **registration** of a new product will be based. They are sometimes split into so-called **Phase IIIa studies** and **Phase IIIb studies**

Phase IIIa study   this term is not often used; the term **Phase III study** is usually adequate. ⇔ **Phase IIIb study**

Phase IIIb study   when a product already has a **marketing authorisation** but the **indication** is being expanded, new **Phase III studies** are needed to demonstrate efficacy in the new indication. Since the Phase III studies in the drug's development have already been completed, these new studies are sometimes referred to as Phase IIIb

Phase IV study   these are studies carried out after **registration** of a **product**. They are often for marketing purposes as well as to gain broader experience with using the new product. ⇒ **post marketing surveillance study, seeding study**

phase of study   denotes different stages within a study. For example, a study may have a **washout period, treatment period, follow-up period**, etc.; each of these could be referred to as a phase rather than a 'period'

phlebotomy   the taking of blood

physician   a medically qualified person who can treat patients. ⇒ **investigator**

physiology   the study of the functioning of the body and body systems

phytomedicine   drugs developed from plants. ⇔ **biologic, pharmaceutical**

pi (π)   a mathematical constant. It is the ratio of the circumference of a circle to its diameter although it has many uses in mathematics and statistics beyond this

pie chart   a circular graph used for showing **percentages**. Schematically it resembles a pie (or a cake) with slices cut out; each slice being proportional (in area) to the proportion of data being represented (Figure 25). ⇔ **stacked bar chart**

pilot   ≈ **pilot study**

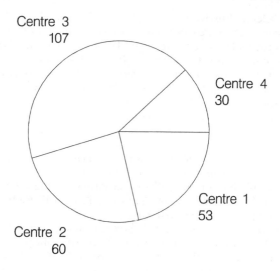

Centre 3
107

Centre 4
30

Centre 1
53

Centre 2
60

Figure 25   Pie chart. The proportion of patients recruited by each of four study centres is represented. In this example, the actual number of patients recruited at each centre is also indicated

pilot study   a small study for helping to design a further, **confirmatory study**. The main uses of pilot studies are to test practical arrangements (for example, how long do various activities take? is it possible to do all the things we want to?), to test questionnaires (do the subjects understand the questions in the way we intended?) and to investigate variability in data. ⇨ **internal pilot study**

pilot test   usually an informal type of **pilot study**

pivotal   something on which a major decision (possibly to continue or cease developing a compound) will depend

pivotal study   a study that is **pivotal**. It may be pivotal for internal company use or for regulatory use. In the latter case, ⇨ **confirmatory study**

placebo   an inert substance usually prepared to look as similar to the active product being investigated in a study as possible. In some situations, the term **vehicle** is used instead. ⇨ **blinding**

placebo control   giving **placebo** to a **control group** of subjects

placebo controlled   ≈ **placebo controlled study**

placebo controlled study a description of a study that implies there is a **control group** who receive **placebo**

placebo effect a nonspecific term used to encompass any (usually beneficial) changes that occur within a group 'treated' with placebo. ⇨ **effect, treatment effect**

placebo group the group of subjects assigned to receive **placebo**. ⇔ **treatment group**

placebo lead in period ≈ **placebo run in period**

placebo period a **period** within a study where subjects are given **placebo**. ⇨ **placebo run in period, placebo washout period**

placebo run in period a **run in period** where all subjects are given **placebo**. ⇨ **placebo washout period**

placebo subject a subject who has been **allocated** to receive **placebo**

placebo treatment an alternative term simply for **placebo**. Strictly, placebo is not a 'treatment' but the term is still commonly used

placebo washout giving subjects **placebo** when the purpose is to allow any other medication that may be in the body to be eliminated. This may be at the beginning of a study (≈ **placebo run in period**) or between periods in a **crossover study**

placebo washout period the time during which **placebo washout** takes place

plagiarise to extensively copy someone else's ideas or work without adequately acknowledging them

plagiarism the act of copying someone else's ideas or work without adequately acknowledging them

plasma the liquid part of blood

platform often used to refer to types (and manufacturers) of different computers. See, for example, **mainframe computer, personal computer**

plausibility check an **edit check** to test if data items appear plausible. This may be based on a simple **range check** or may be a more complex **consistency check**. Note that data that pass a plausibility check may still not be correct

play-the-winner rule a method of assigning treatment to subjects. When the response is **binary**, the next subject will be given the same treatment as the last subject, if the last subject showed a positive response. However, if the last subject showed a negative response, then the next subject will be given the alternative treatment

play-the-winner treatment assignment a method of treatment assignment that uses a **play-the-winner rule**

plot a graph; or to draw a graph

plus and minus ≈ **plus or minus**. Note the distinction between 'and' and 'or' is poorly used. ⇨ **and/or**

plus infinity   the term **infinity** strictly means plus infinity but in some situations it is helpful to distinguish from **minus infinity**

plus or minus ($\pm$)   a term used to describe a range of values, often in the context of a **confidence interval**. For example, if the percentage of subjects responding to a treatment was 35% and 2 **standard errors** for this response rate equalled 5%, then the estimate might be described as '35%, plus or minus 5%'. The term is often (badly and ambiguously) used to refer to either the **standard deviation** or the **standard error** of an estimate

Pocock rule   one of the well known but lesser used **stopping rules** in **group sequential studies**. ⇨ **O'Brien and Flemming rule**

point estimate   a single value of an estimate of a **parameter**. ⇔ **interval estimate**

point estimation   the process of determining a **point estimate**. ⇔ **interval estimation**

point prevalence   the **prevalence** (number of **cases**) of an event at a particular moment in time. ⇔ **period prevalence**

poison   any substance that can cause damage or injury to the body (including one which may cause death)

Poisson data   data that come from a **Poisson distribution**

Poisson distribution   the **probability distribution** for the number of independent events that occur in a fixed period of time

Poisson process   any process (biological, mechanical, etc.) that results in data from a **Poisson distribution**

Poisson variable   a variable from a **Poisson distribution**

polychotomous data   **categorical data** where there are more than two categories but where there is no natural ordering to the categories. ⇔ **ordinal data**

polychotomous variable   a variable that can take one of several (unordered) values

polygon   any shape whose corners are joined by straight lines. ⇨ **frequency polygon**

polynomial   referring to powers of a variable, for example x^2, x^3, $x^{5.1}$

polynomial regression   the use of **polynomials** as **predictor variables** in a **regression model**

polytomous regression   an extension of the methods of **logistic regression** where the **response variable** is **unordered categorical data**, instead of **binary data**. ⇨ **ordered logistic regression**

pool   to gather together from different sources

pooled estimate   an estimate of a **parameter** that uses information from different sources (often different groups of subjects or different studies). It is quite common to use a pooled estimate of variance in **analysis of variance, *t* tests**, etc.

population   the entire group of subjects that could (in theory) be included in a sample. Often the population cannot be exactly defined since it consists of an infinite number of people, but it can be defined in a theoretical way (such as 'all people with raised systolic blood pressure').
⇨ **finite population**

population controls   a **control group** that is not chosen by **randomisation** but that is simply the general population of nontreated patients.
⇨ **historical control, concurrent control**

population demographic fraction   the proportion of the population with some particular **demographic variable**. ⇔ **sample demographic fraction**

population mean   the (unknown) value of the **mean** of a variable in an entire population, usually denoted μ. ⇔ **sample mean**

population parameter   ≈ **parameter**

population standard deviation   the (unknown) value of the **standard deviation** of a variable in an entire population, usually denoted σ.
⇔ **sample standard deviation**

population variance   the (unknown) value of the variance of a variable in an entire population, usually denoted σ^2. ⇔ **sample variance**

portrait   a **page layout** that is higher than it is wide, as in how (most) portrait pictures would be viewed. Portrait A4 paper is 210 mm wide and 297 mm high. ⇔ **landscape**

positive control   ≈ **active control**

positive correlation   **correlation** between two variables such that, as one increases, the other also tends to increase. ⇔ **negative correlation**

positive effect   often used as a synonym simply for **effect**. Sometimes used to refer to any **beneficial effect**

positive predictive value   in a **diagnostic test**, the probability that a person with a **positive result** does actually have the disease (a correct result).
⇨ **negative predictive value, sensitivity, specificity**

positive relationship   an informal term for **positive correlation**. ⇔ **negative relationship**

positive response   a beneficial response to treatment. ⇔ **negative response**

positive result   a result greater than zero. The result of a **positive study**.
⇔ **negative result**

positive skew   describes a distribution that has a long right hand **tail** so that the bulk of the values are at the lower end of the scale. ⇔ **negative skew**

positive study a study that rejects the **null hypothesis** or otherwise successfully fulfils its **objectives**. ⇔ **negative study**

positive treatment effect any desirable **treatment effect**. ⇔ **negative treatment effect**

post hoc to decide on some action after some other event has occurred. ⇔ **prespecifiy**

post hoc analysis any analysis where the decision to do that analysis is taken after seeing other results from the study. The term is also used to describe any decisions about analyses that are taken after the blind has been broken. ⇔ **prespecified analysis.** ⇨ **post hoc comparison**

post hoc comparison a comparison (usually in the form of a statistical **significance test**) that was decided upon after the data were collected, and is generally considered to give less convincing evidence than a prespecified comparison. ⇨ **post hoc analysis**

post marketing surveillance monitoring of the use of and response to products after they have been marketed

post marketing surveillance study a study of a product after it has been marketed but usually taken as being within its **marketing authorisation**. It is a **Phase IV study** rather than a **Phase IIIb study**. ⇨ **seeding study**

postassignment anything that occurs after subjects have been **assigned** (≈ **allocate**) to their **treatment regimen**. ⇨ **postrandomisation**

posterior an abbreviation for **posterior distribution**, **posterior odds** or **posterior probability**. ⇔ **prior**

posterior belief a term used in a similarly informal way to **posterior**

posterior distribution in **Bayesian** statistics, this is the **probability distribution** of a **parameter** after combining the prior distribution with the data

posterior odds in **Bayesian** statistics, this is the **odds** for a **parameter** after combining the **prior odds** with the data

posterior probability in **Bayesian** statistics, this is the **probability** of a **parameter** taking a particular value after combining the **prior probability** with the data

postrandomisation anything that occurs after subjects have been randomised to their treatment regimen. ⇔ **prerandomisation**

postrandomisation follow-up visit any visit that occurs after randomisation. ⇔ **prerandomisation visit**

poststratification the practice of combining data (usually measured on a **continuous scale**) into categories (such as age groups) to form **strata**. The data are then generally analysed or presented according to these strata

posttest distribution ≈ **posterior distribution**

posttest odds ≈ **posterior odds**

posttest probability ≈ **posterior probability**

posttreatment follow-up any follow-up that occurs after treatment has been given

posttreatment follow-up period the time period during which **posttreatment follow-up** occurs

postulate to suggest. A **hypothesis**

potency the strength or **concentration** of a drug

potent a drug that has high **potency**. It is one that has its effects (positive or negative) at very low doses

potential bias **bias** that may occur but which is not a certainty. Often it is not known if any bias has occurred (or will occur) and so any potential bias is important because (although it cannot be demonstrated to have occurred) it cannot be demonstrated not to have occurred

power in statistical **significance tests**, the probability that the **null hypothesis** will be **rejected** if it is not true. Also referred to as **Type II error**. ⇔ **significance level** or **Type I error**

power transformation a transformation of the form $y = x^z$. ⇨ **exponential growth**, **polynomial**

practical significance ≈ **clinical significance**

pragmatic being practical and not idealistic

pragmatic study a study that aims to answer the question of whether or not a treatment works in practice, rather than whether it works in some ideal set of circumstances (for which, ≈ **explanatory study**). ⇨ **intention-to-treat**

preassignment anything that occurs before subjects have been assigned to their treatment regimen

precise the extent to which replicated measurements agree with one another. ⇔ **accurate**. ⇨ **intraobserver agreement**

precision the specification of how **precise** a measurement is

preclinical anything that occurs before drugs in development begin to be tested in humans (before the clinical work begins)

preclinical research any form of research that is **preclinical**

preclinical study a study that is **preclinical**. This often implies that it is conducted in animals

predicted value ≈ **fitted value**

prediction interval a range of values that are considered likely for any individual to take for a particular **parameter**. Prediction intervals are usually based on **models** from past data (Figure 26). ⇔ **confidence interval**

prediction limit the value at the lower or upper end of a **prediction interval**. ⇔ **confidence limit**

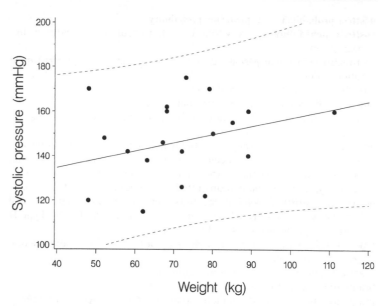

Figure 26 Prediction interval. The regression line predicting systolic blood pressure from weight is subject to uncertainty. The broken lines are a type of 'confidence interval' for predicting an individual's blood pressure, given knowledge of their weight. Note that it is much wider than a true confidence interval, which would show the mean blood pressure for patients of a given weight

predictive value   see **negative predictive value**, **positive predictive value**

predictor variable   ≈ **covariate**. ⇔ **outcome variable**

preference designs   types of study design where an option exists for patients to receive the treatment of their own choice (their own preference) or to be randomised to one or other of the treatments being studied. Examples include **Wennberg's design** and **Zelen's randomised consent design**

preferred term   strict medical terms that should be used to describe signs, symptoms, disease, etc. ⇨ **high level term**, **system organ class**

premature stopping   stopping recruitment to a study before the total planned sample size has been reached. This can occur either formally, in the context of **sequential studies** and **group sequential studies**, or less formally. Possible reasons for premature stopping include sufficient evidence of a treatment effect, sufficient evidence of no treatment effect, cost, slow recruitment, results of other studies being published

preparation   ≈ **product**

prerandomisation   anything that occurs before subjects have been randomised to their treatment regimen

prerandomisation visit   any visit that subjects attend before they have been **randomised**

prescribe   to give a medical treatment

prescription   the written documentation giving details of what a physician has prescribed for a patient

prescription drug   ≈ **prescription only medicine**

prescription event monitoring   a process of keeping records of the numbers of prescriptions issued for a product and the number of adverse events subsequently recorded. ⇨ **post marketing surveillance**

prescription only medicine   a product that can be obtained only with a doctor's prescription. ⇔ **over-the-counter drug**

presentation   the way in which a product is manufactured for patients to use. Common presentations are **tablet, capsule, suspension, pessary, transdermal patch**, etc.

prespecified analysis   any analysis (but in particular related to a statistical **significance test**) that it was agreed would be carried out before the data were collected (or before the **blind** was broken). ⇔ **post hoc analysis**

prespecify   to agree to some action in advance of some other event taking place

pretest distribution   ≈ **prior distribution**

pretest odds   ≈ **prior odds**

pretest probability   ≈ **prior probability**

pretesting   the process of using a statistical **significance test** to help decide what further analysis of data should be conducted. Examples include **stepwise regression**

pretreatment examination   any examination of a subject before treatment is given

prevalence   the number of people in a **population** with a given disease. ⇨ **point prevalence, period prevalence**. ⇔ **incidence**

prevalence rate   the proportion of people in a **population** with a given disease. ⇔ **incidence rate**

prevention study   a study aimed at determining if the **incidence** of a disease can be reduced by preventing people getting the disease. ⇨ **screening study, vaccine study**

preventive medicine   that area of medicine concerned with preventing disease occurring (generally by promoting a healthier lifestyle or healthier living/working environment)

preventive treatment   a treatment used to prevent disease rather than to relieve symptoms. Some drugs can be used for both prevention and treatment, often at different doses

primary   of first importance. ⇔ **secondary**

primary analysis   the most important analysis in a study. Usually an analysis of the **primary endpoint** and the analysis on which the conclusions of the study most heavily rely. ⇔ **secondary analysis**

primary care   the initial care given to a patient immediately after diagnosis. ⇔ **secondary care, tertiary care**

primary care centre   the place where **primary care** is administered. ⇔ **secondary care centre, tertiary care centre**

primary care study   a study carried out on patients with diseases that would generally be treated at a **primary care centre**. ⇔ **secondary care study**

primary data   the most important data for the **primary analysis**. Also used as a synonym for **source data**. ⇔ **secondary data**

primary endpoint   the most important endpoint in a study, providing the **primary data**. ⇔ **secondary endpoint**

primary objective   the most important objective of a study. ⇔ **secondary objective**

primary outcome   ≈ **primary endpoint**

primary prevention   prevention (of disease) at the source where it is likely to be contracted. ⇔ **secondary prevention**

primary prevention study   ≈ **prevention study**

primary result   the most important result of a study. ⇔ **secondary result**

primary variable   the most important variable in a study, used in the **primary analysis** to assess the **primary objective**

principal investigator   the **investigator** who takes overall responsibility for a study (although not responsibility for the wellbeing of patients who are treated by other investigators taking part in the study). ⇨ **subinvestigator**

prior   an abbreviation for **prior distribution**, **prior odds**, or **prior probability**. ⇔ **posterior**

prior belief   a term used in a similarly informal way to **prior**

prior distribution   in **Bayesian** statistics, this is the **probability distribution** of a **parameter** before combining it with the data to obtain the **posterior distribution**

prior odds   in **Bayesian** statistics, this is the **odds** of a **parameter** before combining it with the data to obtain the **posterior odds**

prior probability   in **Bayesian** statistics, this is the **probability** that a **parameter** takes a certain value before combining it with the data to obtain the **posterior probability**

pro drug   a drug that is not effective in its own form but which **metabolises** in the body into a form that is effective

probability   the likelihood, or chance, that an event will occur, often estimated by the **relative frequency**

probability density function   the mathematical function that describes the shape of a **probability distribution**

probability distribution   see, for example, **Normal distribution,** *t* **distribution, chi-squared distribution.** ⟺ **frequency distribution**

probability sample   a sample of people (or objects) where every subject in the population has a known probability of being included in the sample. Note that the probabilities of each subject being included are not necessarily all the same, but they must all be known. ⟹ **random sample, convenience sample, haphazard sample**

probability sampling   the process of taking a **probability sample**

probable error   not a frequently used term but it equals 0.675 of the **standard deviation**. In the case of a **Normal distribution**, the mean minus one probable error equals the **lower quartile** and the mean plus one probable error equals the **upper quartile**

probit analysis   analysis of data in the form of **proportions** after applying a **probit transformation.** ⟹ **logistic regression**

probit transformation   a mathematical function used to transform **proportions** to a scale that ranges from **minus infinity** to **plus infinity.** ⟹ **angular transformation, logistic function**

producer's risk   the **probability** of committing a **Type II error.** ⟺ **consumer's risk, regulator's risk**

product   a pharmaceutical product in the form in which it is presented for use. Bulk product of drug is not usually referred to as a product but **tablets** or **syringes** containing drug are

product licence   ≈ **marketing authorisation**

product licence application   a request submitted to a **regulatory authority** for a licence to market a product for specific **indications**

product life cycle   a term usually used in marketing to describe how sales are likely to change during the time a product is on the market

product monograph   detailed technical information about the chemical makeup of a product. ⟹ **summary of product characteristics**

product-limit estimate   a method for estimating the **survival function** of a set of **survival times** (some of which may be **censored**). ⇨ **log rank test**

prognosis   the probable course of a disease

prognostic   predictive of an outcome

prognostic factor   a **factor** that is predictive of the **outcome variable** in a study. The term factor is often used in this context to include **continuous variables** as well as **discrete variables**. ⇨ **covariate**

prognostic variable   ≈ **prognostic factor**

program   a plan of a sequence of actions to be carried out. With this spelling, the term is generally restricted to mean a **computer program**. ⇨ **programme**

programme   as in **program** but this spelling excludes computer applications. Common uses include a programme of research, a programme of studies, an educational programme

programming   the process of writing **computer programs**

progression   in cancer studies, this is generally regarded as an increase in tumour size of at least 50%. ⇨ **complete response**, **partial response**, **stable disease**

project   a broad term to describe any piece of work that involves more than one simple activity

prone   lying down, face downwards. ⇔ **supine**

propensity score   in studies that do not use **random allocation**, this is a value that indicates (separately for each subject) how likely a subject is to receive any one of the treatments being compared, given a set of **covariates** measured on that subject. ⇨ **outcomes research**, **natural experiment**

proper prior   in **Bayesian** statistics, a **prior distribution** that could be a **probability distribution** in its own right. The most common exception, when a prior distribution is not 'proper', is when a **reference prior** is used that assigns equal probability to all values between **minus infinity** and **plus infinity**

prophylactic   something that prevents something else

prophylactic dose   a dose of a drug that is used to prevent a disease (or symptoms of a disease) from occurring. Usually a prophylactic dose will be much lower than a **therapeutic dose**

prophylactic study   a **prevention study** that uses a drug for prevention, rather than some other means such as change of life style or control of the environment

prophylaxis   prevention

proportion   a fraction, usually in the context of how many subjects experienced some event relative to how many were at risk of experiencing the event. Simple proportions are $\frac{3}{4}$, $\frac{4}{5}$, $\frac{17}{20}$, etc.

proportional   two or more groups having events that occur in the same proportion as each other

proportional hazards   in **survival studies** the case that two **hazard functions** are in the same proportion to each other over the entire time course of a study

proportional hazards regression model $\approx$ **Cox's proportional hazards model**

proportional odds model   a model similar to **Cox's proportional hazards model** but where the **odds** of death (or other specified event) is in the same proportion between two groups over the time course of a study, rather than the **hazard functions** being proportional

proprietary   relating to the owner of a product or information and which is usually confidential, patented, etc.

prospective   something that is planned to take place in the future. $\Leftrightarrow$ **retrospective**

prospective data   data that are planned to be collected as events occur rather than data that were collected for some other purpose but that can be used for a current study

prospective follow-up   **follow-up** of subjects that is planned to take place and which usually does so in a controlled manner. $\Leftrightarrow$ **retrospective follow-up**

prospective study   a study that is planned to take place and which will collect **prospective data**. $\Leftrightarrow$ **retrospective study**, **outcomes research**

prosthesis   a **medical device** to replace bone or joint material

protocol   a written document describing all the important details of how a study will be conducted. It will generally include details of the products being used, a rationale for the study, what procedures will be carried out on subjects in the study, how many subjects will be studied, the design of the study and how the data will be analysed

protocol amendment   a formal written document that describes any changes to be made to a previously written **protocol**

protocol departure $\approx$ **protocol deviation**

protocol deviation   something that happens within a study and that does not fully conform to what was described in the **protocol** (and any **protocol amendments**)

protocol violation $\approx$ **protocol deviation**

protocol violator   a person (usually a subject or patient rather than one of the study staff) who does not fully comply with all aspects of a **protocol** (and any **protocol amendments**)

provocation test $\approx$ **challenge test**

proxy variable   ≈ surrogate

pseudorandom   not completely **random** (although often appearing random)

pseudorandom number   numbers that appear as if they are completely **random** but which are not. Random numbers are very difficult to produce but pseudorandom numbers can be produced in many different ways. They are often used to create the **randomisation code** to decide which subjects receive which treatment or to help in taking **random samples** from large **populations**

psychosomatic effect   physical effects thought to be caused by psychological factors such as stress

psychotropic drug   any drug that has its direct effects on the brain

public health study   ≈ **community study**

publication bias   the situation where there is a tendency for **positive studies** to be more widely published (or otherwise reported) than **negative studies**. This is a particular problem and can cause **bias** in carrying out **overviews** and **meta-analyses**

***P*-value**   the initial 'result' from a statistical **significance test**. It is the probability of getting a result at least as extreme as that observed if the **null hypothesis** is true. It is often misinterpreted as the probability that the null hypothesis is true and for many practical purposes, this may be sufficient. However, that is not the correct interpretation

Q–Q plot ≈ **quantile–quantile plot**

quadratic curved and having squared terms, as in x^2, but no higher order
  polynomials such as x^3 or x^4 (Figure 27)

quadruple blind subjects are **blind** to what medication they receive;
  pharmacists (or those administering treatment) are blind to what
  treatment they are giving each subject; **investigators** (or those assessing
  efficacy and safety) are blind to which treatment a subject was given; and

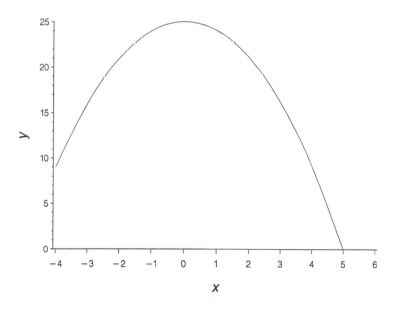

Figure 27 Quadratic. The graph of $y = 25 - x^2$

data management and statistical personnel are blind as to which treatment each subject received. The only use over that of **triple blind** is when the person who gives the treatment to the subject is not the same as the one who subsequently assesses the effect of the treatment. ⇨ **double blind, single blind**

qualitative referring to qualities, rather than quantities⇔ **quantitative**

qualitative data ≈ **categorical data**. ⇔ **quantitative data**

qualitative interaction an **interaction** where the sign of the **treatment effect** changes for different **levels of a factor**. ⇔ **quantitative interaction**

quality a **characteristic** or trait. Note that although **qualitative** excludes **numerical** data, variables such as height, weight and pulse are often referred to as qualities. The term is also, quite differently, used to describe how good something is

quality adjusted life years a term originally developed in cancer studies to balance poor **quality of life** (but possibly with a long life expectancy) with good quality of life (but possibly with shorter life expectancy). ⇨ **person-year**

quality adjusted survival times ≈ **quality adjusted life years**

quality assurance a retrospective assessment of the **quality** of a product or service. ⇔ **quality control**

quality control concurrent procedures to ensure adequate **quality** of a product or service. ⇔ **quality assurance**

quality of life a broad assessment of how a subject feels about his or her life. It includes considering their physical and mental states and their degree of satisfaction against some ideal standard

quality of life measure a measurement scale (usually derived from a questionnaire) to score a subject's **quality of life**

quantal effect a **binary** effect; one that either occurs or does not occur

quantile a set of ranges of data across a **distribution** such that each range includes the same number of observations as every other range. The ranges will not (in general) be of equal widths but they will contain equal numbers of observations. Examples include **deciles, centiles, quartiles**

quantile–quantile plot a plot of the observed **quantiles** of a frequency distribution against those that would be expected under an assumed **probability distribution**, usually that of a **Normal distribution**. If the graph is close to a straight line, then the assumption of a Normal distribution (or other probability distribution) is supported (Figure 28)

quantitative referring to quantities rather than qualities⇔ **qualitative**

quantitative data data that are (usually) measured on a **continuous scale**. ⇔ **qualitative data**

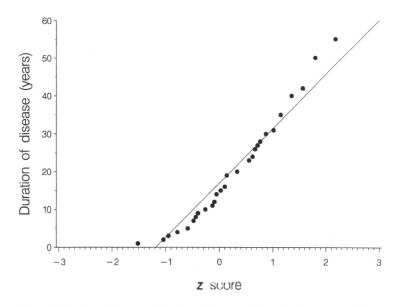

Figure 28 Quantile–quantile plot. The duration of disease is plotted against the z score for a group of 87 patients and the regression line drawn. These data do not fit too well to the straight line, thus indicating a deviation from the Normal distribution. This pattern of lack of fit to a straight line is typical of positively skewed data

quantitative interaction   an **interaction** where the sign of the **treatment effect** remains the same for different **levels of a factor** even though the size of the effect changes. ⇔ **qualitative interaction**

quartile   each of the 25th, 50th and 75th **centiles**. These are also known, respectively, as the **lower quartile**, the **median** and the **upper quartile**

quartile deviation   ≈ **semi-interquartile range**

quasirandom   ≈ **pseudorandom**

quasirandom sample   a sample that is not strictly drawn at **random** but by a **pseudorandom** process

query   a question; usually one that is raised regarding the **validity** of data

questionnaire   a series of related questions

Quetelet's index   an index of obesity (proposed by the Belgian statistician Quetelet). It is calculated as a person's weight (in kilograms) divided by the square of their height (in metres). It is also commonly known as the body-mass index

quick and dirty analysis   an informal term implying that less care is taken with the quality of data and of its analysis than should be, in order to quickly obtain results that are indicative of what the final results will look like

quintile   each of the 20th (20th, 40th, 60th, 80th, 100th) **centiles**. ⇨ **decile**, **quartile**

quotient   the result of dividing one number by another

race ≈ **ethnic origin**

radial plot a graph for plotting **odds ratios** from several different studies. Often used as part of a **meta-analysis** (Figure 29)

radix in **life tables**, the arbitrary number of subjects that are assumed to be alive at day zero

random not happening systematically or predictably

random access a method of storing data in computers such that any of the data can be directly retrieved. An alternative method of storage would involve searching through a magnetic tape until the required data are found (≈ **sequential access**)

random access memory computer memory space that can be accessed by a **random access** method

random allocation the process of randomly deciding which treatments are allocated to which subjects. ⇔ **alternate allocation**

random assignment ≈ **random allocation**

random effect a **categorical variable** where the different levels of the factor are considered to be a random sample of those about which we wish to draw conclusions. ⇨ **random effects model**. Less formally, the term is often used to refer to any effect that is not a **systematic** effect. ⇔ **fixed effect**

random effects model a statistical model that assumes some of the features of the model are randomly chosen from a wider population. Two common assumptions are that subjects are a random selection of all patients with the target disease and that study centres are a random sample of all centres that treat the disease. ⇔ **fixed effects model**

random error ≈ **random variation**. Note that the term 'error' is misleading in that this does not represent mistakes or errors of that kind

random number a number that cannot be predicted with any degree of certainty. ⇨ **pseudorandom number**

random number generator a process (either mechanical or electronic) that produces a series of **random numbers**

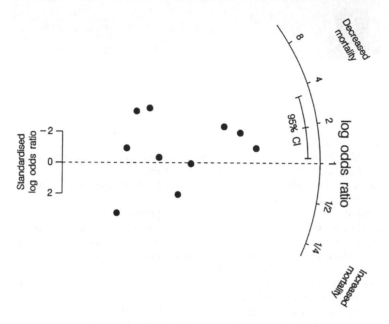

Figure 29 Radial plot. The odds ratios of ten studies of treatment of cardiovascular disease. The more precise studies (those with smaller standard errors of the odds ratio) are plotted further to the right; less precise studies are to the left. The overall summary from a meta-analysis is indicated, along with its 95% confidence interval

random permutation   any random ordering of a given set of items. For example, three treatments could be given in the sequences ABC, ACB, BAC, BCA, CAB, or CBA. ⇨ **randomised block**

random sample   a **sample** from a **population** where every individual in the population has the same probability of being chosen. ⇨ **probability sample**

random sampling   the process of taking a **random sample**

random sampling numbers   **random numbers** used for **random sampling**

random sequence   ≈ **random permutation**

random variable   a variable whose values come from a recognisable **probability distribution** such as the **Normal distribution**, *t* **distribution**, etc.

random variation   variation in measurements that occurs at random and is not predictable. ⇔ **systematic variation**

randomisation   the process of randomising a set of data values, subjects, treatments, etc.

randomisation code   either the entire **random allocation** sequence for all subjects in a study (as in **randomisation list**) or the individual allocation for a single subject

randomisation envelope   a sealed envelope with the **randomisation code** for a subject inside. To maintain **blinding**, it should be opened only in an emergency if it is necessary to know which treatment a subject has been receiving, or at the end of a study so that the analysis can be carried out

randomisation list   a list, produced by a **random** process, that tells which subjects will receive which treatment in a randomised study

randomisation schedule   ≈ **randomisation list**

randomisation test   ≈ **nonparametric test**

randomisation visit   the **visit** at which treatments are randomly assigned to subjects

randomise   to assign a treatment to a subject by **randomisation**

randomised block   a **block** of treatment where the sequence of treatments within the block has been **randomised**

randomised block design   a study design where treatments are packed in **randomised blocks**. ⇔ **completely randomised design**

randomised clinical trial   a clinical trial that uses **randomisation** to decide which subjects receive which treatment. It is often assumed that clinical trials will include randomisation but that is not necessarily the case. This term emphasises the use of randomisation

randomised consent design   ≈ **Zelen's randomised consent design**

randomised control   a subject who has been assigned (by a random process) to the **control group**

randomised control group   the group of subjects who have been assigned (at random) to receive the **control treatment**

randomised group   any **strata** that are formed as a result of **random allocation**

randomised provider design   a study when some subjects are **randomised** to be treated by one group of doctors (who always use the **active treatment**) and other patients are randomised to be treated by a different group of doctors (who always use the **control treatment**)

randomised provider study   a study that is designed as a **randomised provider design**

randomness   the extent to which a process is **random**

range   the difference between the largest and smallest numbers in a set of data

range check   an **edit check** to identify any data values that fall outside a specified lower limit and upper limit

range of distribution   ≈ **range**

range of equivalence   the extent to which two treatments may reasonably be considered to show equivalent efficacy. For example, if one treatment cures 70% of patients, then a range of equivalence might be judged to be from 65% to 75%. Any such range will usually require an element of clinical judgement. ⇨ **bioequivalent**

rank   the order in which data values are placed, either in importance or in size. To place in order of size or importance

rank correlation   a **nonparametric** form of **correlation**. ⇨ **Kendall's tau, Spearman's rho**

rank correlation coefficient   the **nonparametric** statistical measure of **correlation**

rank data   ≈ **ordinal data**

rank order   the numerical order (1st, 2nd, 3rd, etc.) of a set of data values that have been **ranked**

rank order statistic   any **statistic** (any 'function of the data') that is based only on the **ranks** of the data, not their actual values

rank test   ≈ **nonparametric test**

ranked data   data that have been put into **rank order**

rankit   ≈ **probit**

rate   the number of events in a specified time period divided by the number of people that could have had events. ⇔ **proportion**

rate difference   the difference between two **rates**

rate ratio   the ratio of two **rates**. ⇨ **odds ratio, risk ratio**

ratio data   **continuous data** that has a fixed zero. Temperature, for example, does not have a fixed zero point (it depends on which units you use); blood pressure does have a fixed zero (whatever units you use)

ratio scale   the scale on which **ratio data** are measured

ratio variable   a variable measured on a **ratio scale**

raw data   data as recorded in their original, most basic, form. ⇨ **source data** ⇔ **derived variable**

raw score   ≈ **raw data**

reaction   an **event** caused by another event. **Causality** is important and implied. ⇨ **adverse reaction**

rebound effect   a sudden and substantial worsening of a disease that sometimes results when a medication is stopped. The effect may be so large as to make symptoms worse than they were before treatment was initially given

recall   to retrieve; either as to remember or to retrieve documents from an archive or other source of storage

recall bias   any bias in remembering events. Often unpleasant events are easier to remember than pleasant ones, or events that we believe may be important are easier to remember than those we believe to be unimportant

receiver operating characteristic curve (ROC curve)   in a **diagnostic test**, a plot of the **sensitivity** of the test versus one minus the **specificity** of the test

reciprocal   'one over', as in the reciprocal of 7 is $\frac{1}{7}$

recode   to **code** a second time, either to correct errors or because a different set of rules for coding are to be applied

record   a set of data values relating to one individual. To keep a set of data values (by writing them on paper, storing them on a computer, etc.)

record linkage   the process of ensuring that different **records**, kept by different people, for different reasons but all relating to the same individual can all be identified as relating to the same individual (and so are 'linked' together)

recruit   to **enrol** (usually subjects into a study)

recruitment   the process of **enrolling** subjects into a study. Also used to refer to the number of subjects that have been enrolled

recruitment log   a **log** kept of who has been recruited into a study. ⇔ **screening log**

recruitment period   the time during which recruitment of subjects into a study takes place

recruitment rate   the number of subjects recruited into a study divided by the time it has taken to recruit them. For example, 'three patients per week'

recruitment target   the number of subjects that it is planned to recruit into a study. The planned **sample size**

rectangular distribution   ≈ **uniform distribution**

refer   to indicate that a patient should go to a more senior expert or **secondary care centre** for medical treatment

reference bias   a tendency to quote some sources of reference material in preference to others, particularly when those that are quoted support one's own point of view more than those which are not cited. Important in **meta-analysis** and **overviews**

reference group   ≈ **control group**

reference interval   ≈ **reference range**

reference limits   the lower limit and upper limit of a **reference range**

reference population   the population from which a sample has been taken

reference prior   ≈ in **Bayesian** statistics, a **prior distribution** that assigns equal probability to all values for a **parameter** between **minus infinity**

and **plus infinity**. ⇨ **uniform prior**

reference range   a range of values for a variable that encompasses most of the data that would be expected from the subjects within the context that the data are being collected. This can therefore be used to help identify abnormal values and **outliers**. ⇔ **normal range**

reference value   a value from which others are measured. This may be 'time zero' (as the beginning of the study, not the beginning of all time) or a **baseline** measurement, from which changes are measured

referral centre   a hospital or other health care centre that can supply more expertise than a **primary care centre**; patients may thus be referred here. ⇨ **secondary care centre**

regimen   a set of procedures. ⇨ **treatment regimen**

region   an area (either geographical or of a graph)

region of rejection   ≈ **rejection region**

register   to record the fact that someone is taking part in a study. ⇨ **recruitment log, screening log**. To obtain a **marketing authorisation** for a new product

registration   the process of **registering** a new product

registration dossier   a large set of documents sent to a **regulatory authority** containing all the known safety, efficacy and manufacturing data about a new product. The purpose of the dossier is to support an application for a **marketing authorisation**

registration phase   any work carried out before **registration** of a product, particularly **Phase III studies**

registration study   ≈ **Phase III study**

registry   a list or **log**

regress   to move backwards (particularly with respect to progress). To use statistical **regression** methods

regression   the statistical term for building models that try to predict an **outcome variable** from one or more **covariates** (Figure 30)

regression analysis   the process of using **regression** methods

regression coefficient   a **coefficient** in a **regression model**. In simple **linear regression**, this is the same as the slope of the line on a graph

regression curve   ≈ **regression line**

regression diagnostics   methods to determine if a **regression line** is a good fit to data or not

regression discontinuity design   an extreme form of **cutoff design** that has no middle group. All subjects with **baseline** scores below some **cutoff** value are assigned to one treatment group; all subjects with scores above that cutoff are assigned to the alternative group

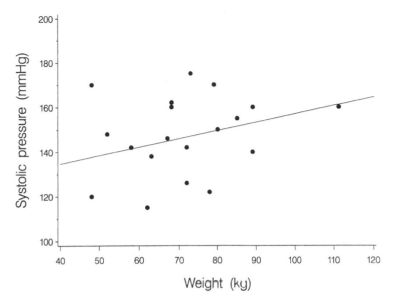

Figure 30 Regression. A simple relationship between systolic blood pressure and subjects' weight. In this example it is clear that as weight increases so too (on average) does blood pressure

regression equation the equation of a **regression line**, or **regression surface**

regression line the line that is the best statistical **model** to describe a **response variable** from a single **covariate**. ⇨ **regression surface**

regression model a general term for the equation of either a **regression line** or a **regression surface**

regression surface in **multiple regression**, instead of producing a **regression line** through the data, the equation of a surface or plane through the 3-, 4-, or n-dimensional space is produced instead

regression through the origin a **regression model** where the **regression line** is forced to go through the **origin** of the graph (the point where $x = 0$ and $y = 0$)

regression to the mean the tendency that, when a variable is measured more than once, if it is very extreme compared with the rest of the distribution of values on its first measurement, it is more likely to be

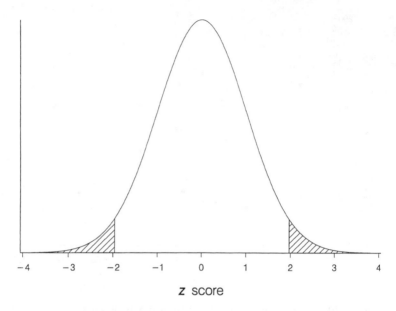

Z score

Figure 31   Rejection region. This example is from the Normal distribution and is used with a *z* test (or with large samples in a *t* test). If the vale of *z* is less than −1.96 or greater than +1.96 (the two shaded rejection regions) then the *P*-value is less than 0.05

closer to the mean on subsequent measurements

regressor   ≈ **covariate**

regular   occurring at very similar time intervals. 'Regular' should not be confused with 'frequent'. An event may occur regularly but infrequently (say once every year) or frequently but irregularly (say once every 5–10 minutes)

regulator's risk   the **probability** of committing a **Type I error**. ⇨ **consumer's risk**. ⇔ **producer's risk**

regulatory agency   ≈ **regulatory authority**

regulatory authority   government formed agencies who are empowered to make decisions concerning whether or not a new product should be given a **marketing authorisation**

regulatory requirement   something that a **regulatory authority** insists is done

reimburse   to give back money to someone who has paid for something. Typically used in the sense of whether an agency (government or otherwise) will pay for the products that a doctor prescribes

reject   to turn down or to declare as untrue. Used particularly in the context of rejecting the **null hypothesis** in statistical **significance tests**

rejection   a declaration that something is untrue

rejection error $\approx$ **Type I error**

rejection region   in statistical **significance testing**, values of the **test statistic** that will lead to rejecting the **null hypothesis**. For the most common type of *t* **test** comparing two means at the 5% significance level, the rejection regions are from **minus infinity** to -1.96 and from $+1.96$ to **plus infinity** (Figure 31)

relapse   the reoccurrence of signs or symptoms of a disease that had been suppressed

relational database   data held on a computer in a set of different tables (each with rows and columns), where the tables can be looked at individually or separately but where there is a key to link related data from different tables

relationship   when two (or more) variables are related in some way. $\Rightarrow$ **correlation**

relative change   the **absolute change** in one treatment group expressed as a proportion or percentage of the absolute change in another treatment group

relative efficacy   the amount of **beneficial effect** produced by one product over and above that produced by another (or a placebo), usually expressed as a percentage. This is essentially the same meaning as the strict interpretation of **treatment effect** but is perhaps a more forceful way of stating it

relative efficiency   the efficiency (*sic*) of one type of study design compared with another. In many cases, efficiency is measured in terms of the reciprocal of the **variance** of the treatment effect. So a design that gives a small variance has a higher efficiency than one with a large variance

relative frequency   the number of times an event occurs divided by the total number of events that occurred (so expressed as a fraction)

relative frequency distribution   the number of times each of several events occurs, divided by the total number of events occurring. $\Leftrightarrow$ **frequency distribution**

relative odds $\approx$ **odds ratio**

relative risk $\approx$ **risk ratio**

reliability   the extent to which the same variable measured on more than one occasion gives similar results. ⇨ **intraobserver agreement**

reliable a variable measured on more than one occasion that gives similar results each time

remote data entry this term would seem to mean **data entry** that is done away from the place where subjects are studied. However, data entry typically takes place at a **data centre** and from the perspective of those who work at a data centre; 'remote' data entry implies data entry at the site where subjects are studied

renal metabolism **metabolism** (of drug) through the kidneys. ⇨ **hepatic metabolism, pharmacokinetics**

repeat observation to read and record a data value more than once, usually with the aim of increasing **precision**. ⇔ **repeated measurements**

repeat reading ≈ **repeat observation**

repeatability the extent to which a process or measurement is **reliable**

repeated dose design a study where each subject receives more than one dose of a drug (or placebo). ⇔ **single dose design**

repeated dose pharmacokinetic study a study to investigate the **pharmacokinetic** effects of a drug given in repeated doses. ⇔ **single dose pharmacokinetic study**

repeated dose study a study to investigate the effects of a drug given in repeated doses. ⇔ **single dose study**. ⇨ **repeated dose pharmacokinetic study, repeated dose toxicity study**

repeated dose toxicity study a study to investigate the **toxicity** of a drug given in repeated doses. ⇔ **single dose toxicity study**

repeated measurements more than one measurement of the same variable on the same subject but taken at different times. ⇔ **repeat observation**

repeated measurements analysis of variance **analysis of variance** methods used for analysing **repeated measurements**

repeated measurements design a study in which subjects have several measurements of the same variable taken at different times. ⇨ **longitudinal study, crossover design**

repeated measures analysis this is generally taken as an abbreviation of **repeated measurements analysis of variance** but should refer more broadly to any sort of analysis of **repeated measurements**

replicate to reproduce someone else's work (to confirm or refute their findings) or to make **repeat observations** (usually to gain precision)

replicate observation ≈ **repeat observation**

report a document describing (usually) the objectives, results and conclusions from a study. Sometimes it may contain only results

representative   typical; **unbiased**

representative sample   ≈ **random sample**. Note that the term is sometimes used to describe a sample that has not been taken at **random** but which is thought to be similar to one that would have resulted from random sampling

reproduce   ≈ **replicate**

reproducibility   ≈ **repeatability**

reproductive and developmental toxicity study   a study of the **toxicity** of a drug with special reference to the reproductive organs

rescue medication   medication used to control signs and symptoms of a disease, particularly when a patient is taking another medication for **prophylaxis**

rescue treatment   ≈ **rescue medication**

research   careful and thorough investigation to discover new facts

research design   the **design** (particularly of an **experiment**) for the purpose of **research**

research ethics   ethical behaviour related to the way that **research** is carried out. Often this highlights the difference between **individual ethics** and **collective ethics**

research ethics committee   a group of people (with a variety of technical, scientific and medical qualifications, as well as lay persons) who review research proposals to ensure they conform to accepted standards of ethics. ⇨ **local research ethics committee**, **multicentre research ethics committee**

research group   a group of people involved in the same **research project** or in similar research projects

research hypothesis   ≈ **alternative hypothesis**

research project   the topic that is being researched

research proposal   a plan of how a **research project** will be run. Such proposals are often produced in order to obtain funding for the project. It is usually an earlier document than a study **protocol**

research question   ≈ **alternative hypothesis**

research subject   a person (patient or not) who takes part in a **research project** (voluntarily or not). ⇨ **guinea pig**

researcher   one who carries out **research**

residual   the difference between the **observed value** of a variable and the value predicted from a statistical **model** (Figure 32)

residual mean square   ≈ **residual variance**

residual sum of squares   the sum of all the squared (that is, x^2) **residuals**. Used extensively in **analysis of variance** and **regression**

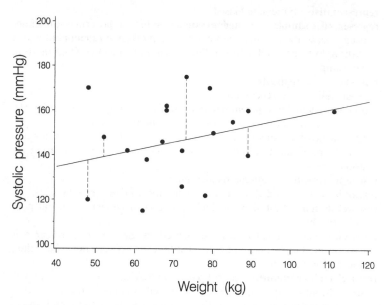

Figure 32 Residual. For this simple regression example, four of the residuals (the vertical distances of each point from the regression line) are indicated

residual variance   any unexplained **variance** or **variation**
respond   to show an effect, usually a **positive effect** (or **beneficial effect**) after treatment
responder   someone who **responds**
response   the reaction of a subject to a treatment. This includes physical responses, psychological responses, **positive responses**, **negative responses**, etc.
response bias   any **bias** that is caused by a systematic difference between those people who respond (typically to a questionnaire) and those who do not respond. Often, those with 'extreme' opinions may be more likely to respond than those with mediocre opinions
response rate   the number of subjects who respond divided by the number of subjects who could have responded. The **proportion** of subjects who respond

response surface   ≈ **regression surface**

response variable   in a **regression model**, the variable that is to be predicted from the **covariate**(s). Also called the **dependent variable**

restricted randomisation   a method of treatment assignment that is mostly **random** but has some restrictions. The most common restrictions are that the total sample size in each group will be the same (or at least fixed), and that within each **block** there will be a fixed number of each of the treatments. ⇔ **completely randomised design**

result   a finding or an observation. What arises from a series of mathematical calculations

retrospective   something that occurred in the past. ⇔ **prospective**

retrospective follow-up   to get follow-up data after any events have occurred rather than to collect it as the events occur. ⇔ **prospective follow-up**

retrospective study   a study that uses already available data rather than one that sets out (**prospectively**) to collect data. ⇔ **prospective study**

review   to check for **quality** or to survey the literature to find out known information on a subject. ⇨ **overview**

review group   a group of people who are responsible for reviewing literature on a subject

rhythm   a regular cycle of events. ⇨ **circadian rhythm, cyclic variation**

ridit analysis   a method of analysis for **ordinal data** when those data are assumed to be from an underlying (but unmeasurable) **continuous variable**

right censored   when measuring the time when an event occurs, the events that do not occur within the study follow-up period are right censored. ⇔ **left censored**

right censored data   when the time that an **event** happens is not known but it is known that the time is at least some amount. ⇔ **left censored data**

right censored observation   ≈ **right censored data**

right skew   ≈ **positive skew**

right tail   the values in a **distribution** that are large (typically meaning those that are greater than the **mode**). ⇔ **left tail**

right–left design   a design where subjects use one medication on the left side of their body and another medication on the right side of their body. The analysis can then be made by **paired comparisons**

right–left study   a study that is designed as a **right–left design**

risk   the **probability** of an event occurring. Note that the consequence of an event occurring may be much greater or lesser than the risk: the two terms are distinct

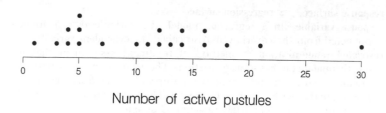

Number of active pustules

Figure 33 Rug plot. The number of active pustules for each of 20 subjects with facial acne. With discrete data, to improve clarity, it is often helpful to offset equal values a little. This then begins to show some similarity to a bar chart

risk assessment   the assessment of the size (and usually consequences) of a **risk**
risk–benefit ratio   the relative weighting of the **risk** (and consequence) of an action if it fails to the benefits gained from that action if it succeeds. ⇨ **therapeutic ratio**
risk difference   the difference between two **risks**. ⇔ **risk ratio**
risk factor   ≈ **prognostic factor**
risk profile   the set of variables that help to define the size of a **risk**. ⇨ **propensity score**
risk ratio   the ratio of two **risks**. ⇨ **odds ratio, rate ratio**
risk set   the subjects in a study at any point in time (that is, all those who are at risk of an event occurring)
robust   stable under a variety of circumstances. This might apply to a chemical that is stable under a variety of environmental conditions (≈ **shelf life**) or, for statistical interpretations ≈ **robust estimator**
robust estimator   a method of estimating a **parameter** that is **robust** to different assumptions about the data and to including or excluding certain of the data values
root   abbreviation of square root
root mean square   ≈ **root mean square error**
root mean square error   the square root of the **mean square error**; the **standard deviation**. ⇨ **analysis of variance**
Rosenthal effect   subconscious bias that tends to lead investigators to see the effects in data that they are looking for. This is one particular reason for **blinding**

round   to approximate, usually this reduces the **accuracy** of a measurement by removing some of the decimal places. ⇨ **truncate**

round off   ≈ **round**

rounding error   bias or error introduced to data when they are rounded

route of administration   the method by which drug is delivered to the body. This may be orally, by injection, through the skin by a transdermal patch, etc.

row vector   see **vector**

rug plot   a graphical display of a **continuous variable** on a horizontal line (Figure 33)

rule   a set way of doing things. ⇨ **decision tree**

rule of diminishing returns   ≈ **eighty–twenty rule**

run in   to enter a study but not be given any study medication immediately. A useful way of gaining time to check **inclusion criteria** and prepare subjects for the practical aspects of the study. ⇨ **washout**

run in period   the time during which the **run in** occurs. ⇨ **washout period**

S shaped curve ≈ **sigmoid**

safe having zero risk (or an acceptably low level of risk)

safe and effective a treatment that is without risk and which works. More usually the term implies that the **risk benefit ratio** is favourable

safety relating to how **safe** something is

safety committee ≈ **data and safety monitoring committee**

safety data any data relating to the safety of a product. Usually this refers to data on **adverse events** and **laboratory data** (haematology, biochemistry, etc.)

safety data monitoring committee similar to a **data and safety monitoring committee** but one with the sole purpose of monitoring the safety data in a study

safety margin an extra safeguard introduced so that, although a product is not as safe as was believed, because we have required it to be even safer than was necessary its actual safety will be within the limits that we require. ⇨ **conservative**

safety monitoring the process of **monitoring** data to ensure that a product is safe. ⇨ **data and safety monitoring committee**

safety population the sample of subjects recruited into a study who are used to assess the **safety** of the products being compared. ⇨ **intention-to-treat population, per protocol population**

safety report a report on the safety aspects of a product

safety review a review of all the **safety data** on a product (this would generally include consideration of **adverse events** and **laboratory data**)

safety study a study primarily intended to assess **safety**, rather than to assess **efficacy**

safety variable a characteristic of a subject that results in **safety data**. ⇔ **efficacy variable**

sample a **subgroup** of a **population**, usually intended to be representative of the population. ⇨ **random sample**

sample demographic fraction the proportion of subjects in a sample with some particular **demographic** characteristic. ⇔ **population demographic fraction**

sample mean   the mean of a sample of data (used to estimate the **population mean**), usually denoted $\bar{x}$. ⇔ **population mean**

sample size   the size of a **sample**: either the number of subjects that it is intended to recruit or the number of subjects that actually were recruited

sample size calculation   a calculation to determine how many subjects are needed in a sample to meet the **objectives** of the study. ⇒ **Type I error, Type II error**

sample size requirement   the required number of subjects needed to meet the **objectives** of a study, arrived at either by a **sample size calculation** or by some other means

sample standard deviation   the **standard deviation** of a sample of data (used to estimate the **population standard deviation**). It is usually denoted s

sample statistic   ≈ **statistic**. This term emphasises that a statistic must (by definition) be based on a **sample**, rather than the **population**. ⇔ **parameter**

sample variance   the **variance** of a sample of data (used to estimate the **population variance**), usually denoted s^2. ⇔ **population variance**

sampling   the act of taking a **sample** (or samples) from a population

sampling distribution   the **probability distribution** of a **test statistic**. For example the sampling distribution of the mean is a **Normal distribution**

sampling error   ≈ **sampling variation**

sampling frame   a description of the **population** from which a **sample** is to be taken. With **infinite** sized populations it is not possible to explicitly define the sampling frame

sampling method   the way in which a sample is taken, for example **random sampling, probability sampling**, etc.

sampling variation   uncertainty (rather than pure error) caused because we have taken a **sample** rather than the entire **population**. In general, every sample will give different **estimates** of the **parameters**

sampling with replacement   when a subject has been selected from a **population**, the relevant data are recorded and, in this situation, that subject is eligible to be sampled again (they are put back into the population that is being sampled). ⇔ **sampling without replacement**

sampling without replacement   when a subject has been selected from a **population**, the relevant data are recorded and, in this situation, that subject becomes ineligible to be sampled again (they are not put back into the population that is being sampled). ⇔ **sampling with replacement**

SAS®   a computer package widely used for data management and statistical analysis and reporting

saturated model   a statistical model that has the same number of **parameters** fitted as there are total **degrees of freedom**. The consequence of this is that there are no degrees of freedom for estimating the **residual variance** so that the **goodness of fit** of the model cannot be assessed. ⇨ **parsimony**

scale   ≈ **measurement scale**

scaled deviance   a term used in **generalised linear models** to describe the amount of variation or **dispersion** in the data

scatter diagram   ≈ **scatter plot**

scatter plot   a graph showing **bivariate data** on an x axis and a y axis (Figure 34)

scattergram   ≈ **scatter plot**

scedasticity   the degree to which the **variances** of data values are equal. ⇨ **homoscedastic, heteroscedastic**

sceptical prior   in **Bayesian** statistics this is a **prior distribution** that reflects scepticism or pessimism

schedule   a sequence of events or activities and when they are to occur. ⇨ **protocol**

scheduled visit   a **visit** that is planned for in a **protocol**

Scheffé's test   a statistical **multiple comparison test** that ensures the **Type I error** rate is controlled. ⇨ **experimentwise error rate, comparisonwise error rate**

scientific hypothesis   an uncommon term for the **alternative hypothesis**

scientific method   the principles of how to carry out objective, unbiased, efficient research through experiments and learning from data and results

scientific misconduct   ≈ **fraud**

scientist   a person who is trained and works according to the **scientific method**

screen   to check if a subject fulfils certain criteria. Often the criteria being considered are the **inclusion criteria** and **exclusion criteria** for a study; the term also applies to screening for a disease. ⇨ **screening test**

screening log   a **log**, or record, that is kept of all subjects that are **screened** for possible inclusion into a study (whether they are subsequently included or not). ⇔ **recruitment log**

screening programme   a widespread plan to carry out screening for a particular disease or diseases

screening study   a study of the value or accuracy of a **screening test**

screening test   a relatively quick and inexpensive test to determine if a

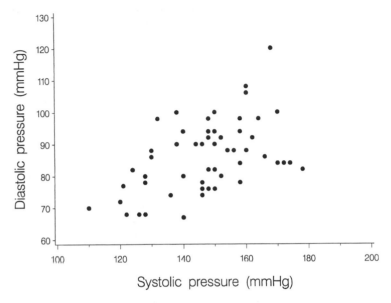

Figure 34   Scatter plot. When two variables are plotted in this way, whether to draw a regression line or not, the basic graph is called a scatter plot. The *x* axis is the horizontal axis (in this case systolic blood pressure), the *y* axis is the vertical axis (in this case diastolic pressure). If the general slope is upwards, the relationship is often called a direct relationship, or a positive relationship

person has a particular, or any of several, diseases. ⇔ **diagnostic test**
screening visit   a visit (typically the first visit) used to complete the **screening log** and to decide if a subject is eligible to enter a study
second order interaction   ≈ **three factor interaction**
secondary   not **primary**, but the term is generally used to include many things that are not primary, including those that are of third, fourth, etc. importance
secondary analysis   analysis (or analyses) that are of secondary importance in a study. ⇔ **primary analysis**
secondary care   care given to a patient after they have been **referred** from a **primary care centre**. ⇔ **primary care**
secondary care centre   the place where **secondary care** is given. A

secondary care centre may have more expertise or resources than a **primary care centre** for treating a particular disease

secondary care study   a study carried out on subjects who would generally be treated at a **secondary care centre**. ⇔ **primary care study**

secondary data   data that are not of primary importance but may be used to answer **secondary objectives** of a study

secondary endpoint   one of (possibly many) less important **endpoints** in a study than the **primary endpoint**

secondary objective   one of (possibly many) less important **objectives** in a study than the **primary objective**

secondary outcome   ≈ **secondary endpoint**

secondary prevention   usually meant as prevention of reoccurrence of a disease that was previously cured. For example, patients who have had a myocardial infarction may be treated for that infarct and then given dietary and lifestyle advice to try to prevent a subsequent infarct. ⇔ **primary prevention**

secondary result   any result of secondary importance in a study. ⇔ **primary result**

secondary variable   the term is broadly used to encompass any variable that is not a **primary variable**

secular   relating to time. ⇨ **temporal**

secular trend   a trend (often **cyclic variation**) in **time series** data that is related to time. ⇨ **circadian rhythm**

seeding study   an outdated term for a **Phase IV study**. Seeding studies used to have very little (if any) scientific merit and were purely marketing exercises disguised as scientific studies

segmented bar chart   ≈ **stacked bar chart**

selection bias   the **bias** caused by the fact that the types of subjects who take part in studies are not a **random sample** of the **population** from which they are drawn. ⇨ **external consistency**

selection method   the method by which subjects are selected to be recruited into studies. Often this is not within the control of the trialists. ⇨ **external consistency, selection bias**

semi-interquartile range   half the difference between the **upper quartile** and the **lower quartile**. ⇨ **interquartile range, probable error**

semiquantitative   usually interpreted as similar to **ordered categorical**. **Quantitative** refers to measurement, semiquantitative refers to measurements that are recorded on a very crude scale. ⇨ **likert scale**

sensitive   capable of measuring very small quantities. Also refers to **personal data** about subjects that they may wish to keep private

sensitive question   a question that asks for private information that a subject may not wish to disclose. Examples include drug usage, sexual activity, etc. ⇨ **personal data**

sensitivity   in a **diagnostic test**, the proportion of subjects with a disease who are correctly identified as having the disease. ⇔ **specificity**

sensitivity analysis   **secondary analyses** carried out by varying the assumptions that are made about the data and models used, including or excluding unusual data points, (**outliers**), etc. The purpose of such analyses is to see if the results and conclusions from a study are **robust**

sequelae   the long term consequences following a disease or an **adverse event**

sequence   a set of items or data values ordered in some logical way. Often (but not necessarily) the ordering is in time. ⇨ **sequential, treatment sequence**

sequential   a sequence in time. ⇨ **time series**

sequential access   in computing terms this refers to the method of accessing data on a physical storage device such as a magnetic tape. ⇔ **direct access**

sequential analysis   special types of statistical analyses that are relevant to **sequential designs**. ⇨ **group sequential analysis**

sequential design   a general type of study design, in which subjects are recruited and the **accumulating data** analysed after every subject has completed the study. The analysis does not wait until a fixed number of subjects have completed the study. The study continues to recruit until a **positive result** or **negative result** becomes evident. ⇨ **closed sequential design, open sequential design, group sequential design**

sequential file   a computer datafile that stores records one after another and is accessed by a **sequential access** method

sequential method   ≈ **sequential analysis**

sequential study   a study that uses a **sequential design**

serendipity   finding useful information by luck, usually when looking for something unrelated

serial   relating to, or arranged in, a **series**

serial correlation   the **correlation** between pairs of adjacent data values in a **time series**. That is the correlation between the first and second values; the second and third values; the third and fourth values, etc. ⇨ **lag, moving average**

serial measurements   ≈ **repeated measurements**

series   a set of items coming one after another (usually in time)

serious adverse event   a regulatory term with a strict meaning. It includes all **adverse events** that result in death, are life threatening, require inpatient hospitalisation or prolongation of existing hospitalisation, result in disability or congenital abnormality. Note that some of these could be nonserious (and quite routine) in a medical sense and they are certainly not necessarily related to study medication

serious adverse experience $\approx$ **serious adverse event**

serious adverse reaction   an **adverse reaction** that is serious by any generally accepted medical standards. $\Rightarrow$ **serious adverse event**

serum   the fluid that forms when blood clots

server   in a computer system with a **local area network** or **wide area network** there is generally a central computer that receives information from each of the remote systems and passes on to other remote systems. This is called the server

sex   male or female

sex ratio   the ratio of males to females (or vice versa) in a study, in a **sample** or in a **population**

sham   an alternative term for a **placebo** but used particularly when the form of the **active treatment** is not a conventional **tablet**, **capsule**, etc. If the 'treatment' under investigation were a certain type of food, then a sham meal might be prepared that looked and tasted the same as the 'test meal' but which lacked certain important vitamins (for example). It would be more usual to call this a sham meal than a placebo meal

sham effect   **placebo effect** when the term **sham** is used in preference to **placebo**

sham procedure   a procedure used as a **sham**

sham treatment   **placebo treatment** when the term **sham** is used in preference to **placebo**. This illustrates the very broad interpretation of the word **treatment**

shelf life   the length of time that a drug (or other product) is expected to last before it becomes unfit for its stated use (assuming it is kept under suitable conditions). Many products can be stored on an open shelf: others may need special conditions such as refrigeration but the term 'shelf life' is still used

shift table   a cross-classification often used for presenting **laboratory data**. The baseline and **end of treatment** values are each categorised as below normal, within the **normal range** (or **reference range**) or above normal. The cross-classification with three rows and three columns in Table 13 shows how many subjects went from low to low, low to normal, low to high, normal to low, etc.

shrinkage   reducing the size of the estimate of an **effect** towards the mean

Table 13   Shift table of serum calcium values from below, within or above the laboratory reference range in 230 patients

Before treatment	After treatment		
	Below	Within	Above
Below	3	11	1
Within	5	186	12
Above	0	7	5

of other effects when it is considered to be unusually large. ⇨ **random effects model**

side effect   usually synonymous with **adverse reaction** but can sometimes be used to refer to a secondary beneficial effect

sigmoid   a curve resembling (slightly) the shape of the letter 'S' (although a rather 'stretched' image of it; Figure 35). It is often useful when analysing **proportions**. ⇨ **logistic curve**

sign   plus (+) for positive numbers or minus (−) for negative numbers.

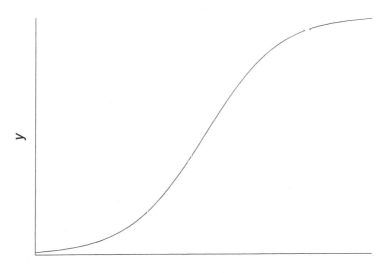

Figure 35   Sigmoid. The broad similarity to a letter 'S' is apparent, albeit with the top and bottom extremes pulled out

Also a feature of a disease that is observable (such as a rash or swelling). For this meaning, ⇔ **symptom**

sign test   a statistical **significance test** for **paired data** that simply considers whether each subject scores higher on one treatment (a positive change) or worse (a negative change) than another. If the **null hypothesis** of no treatment effect is true, we would expect about half of the signs to be positive and half to be negative. ⇨ **McNemar's test, Wilcoxon matched pairs signed rank test**

signal to noise ratio   the ratio of the size of an effect to (some measure of) the size of the variation in the data. ⇨ **coefficient of variation**

signed rank test   ≈ **Wilcoxon matched pairs signed rank test**

significance   see **clinical significance, statistical significance**. The distinction between these two terms is very important but often poorly made

significance level   a prespecified probability at which it is agreed that the **null hypothesis** will be rejected. This is usually set at 0.05 (or 5%), less commonly at 0.01 (or 1%), but any value may be appropriate. ⇨ *P*-value, **strength of evidence**

significance test   a very widely used approach to analysing data. It is a set of methods used to test a **null hypothesis**. Conventionally if the observed data are sufficiently extreme to be unlikely to be consistent with the null hypothesis (that is, they have a small probability of being consistent with the null hypothesis), then that null hypothesis is rejected. ⇨ **hypothesis test**

significant   see **clinically significant, statistically significant**. The distinction between these two terms is very important but often poorly made

significant risk study   a study of a **medical device** that potentially poses an important risk to the subjects who take part. ⇔ **nonsignificant risk study**

simple hypothesis   an **alternative hypothesis** of the form that the **parameter** of interest (possibly the treatment effect) is equal to some specified value. ⇔ **composite hypothesis**

simple random sample   a **random sample**. The term 'simple' is sometimes added to reinforce the fact that there are no restrictions in the sample. ⇨ **stratified random sample**

Simpson's paradox   the case where the relationship between two variables (usually **binary variables)** can be the same for each level of a third (binary) variable; but if the third variable is ignored, then the relationship changes in magnitude (or even in sign). Most frequently quoted in the context of **two-by-two tables** (Table 14)

simulate   to artificially generate **random numbers** that resemble a real problem. The purpose is to solve problems by observing what happens

Table 14   Example of Simpson's paradox. Overall (Table 14a), it appears that Treatment A is worse than Treatment B (68% treatment success vs 83%). However, when the table is stratified by gender it seems that Treatment A is always better (in males (Table 14b), 93% treatment success vs 87%; in females, (Table 14c) 73% treatment success vs 69%).

(a) All patients

	Treatment A	Treatment B
Treatment success	273	289
Treatment failure	77	61

(b) Males

	Treatment A	Treatment B
Treatment success	81	234
Treatment failure	6	36

(c) Females

	Treatment A	Treatment B
Treatment success	192	55
Treatment failure	71	25

in the simulations. ⌐⟩ **Monte Carlo Method**

simulation   the process of **simulating**. ⇨ **Monte Carlo method**

single blind   the case where only one party is **blind** to the treatment allocation. It is usually either the investigator or the subject that is blind but the term on its own does not differentiate. ⇨ **double blind, triple blind, quadruple blind**

single centre study   a study that takes place in one investigational centre (usually with one investigator). ⇔ **multicentre study**

single data entry   in contrast to **double data entry**, data are entered only once; this may then be followed by proof reading

single dose design   a study where each subject receives only one dose of a drug (or placebo). ⇔ **repeated dose design**

single dose pharmacokinetic study   a study to investigate the **pharmacokinetic** effects of a single dose of a drug. ⇔ **repeated dose pharmacokinetic study**

single dose study   a study carried out as a **single dose design**

single dose toxicity study   a **single dose study** whose primary objective is to investigate **toxicity**

single entry   ≈ **single data entry**

single mask ≈ **single blind**

single patient study ≈ **n-of-1 study**

single sample t test a statistical **significance test** to test the **null hypothesis** that the mean of a **population** is equal to some prespecified value. ⇨ **paired t test**

site ≈ **investigational site**

site visit a visit (usually by study monitoring staff) to an **investigational site**

size of a test ≈ **significance level**

skew not symmetric. ⇨ **negative skew, positive skew**

skewed data data that come from a **skewed distribution**

skewed distribution a distribution that is not symmetrical. For example, the distribution shown in Figure 15 (see **histogram**) shows **positive skew**. ⇨ **negative skew**

skewness the degree to which a distribution is not symmetric. ⇨ **negative skew, positive skew**

slope the steepness (usually of a **regression line** or **regression surface**)

small area network ≈ **local area network**

small expected frequencies usually restricted to **expected frequencies** (usually less than five) in **contingency tables**. The expected frequencies are based on the **null hypothesis** of independence between the rows and columns

small sample this term is highly dependent on the context. However, for many statistical analyses, samples of fewer than about 20 or 30 subjects are often referred to as being 'small'; samples of more than 30 subjects may often be referred to as 'large'

smoothing the process of removing small **peaks** or **troughs** in data, usually by **regression** methods

snapshot a recording of data at one particular point in time, without regard to any **secular trends** that may be present

soft data ≈ **subjective data**

soft endpoint ≈ **subjective endpoint**

soft measurement ≈ **subjective measurement**

soft outcome ≈ **subjective outcome**

software all types of computer programs. ⇨ **hardware**

sojourn period the time between a disease being detectable by a **diagnostic test** and development of any symptoms. ⇨ **incubation period**

source data the first place where data are recorded. Often data may be recorded in a patient's hospital notes and also in a **case record form**. In these situations, the hospital notes would be considered as the source data

source data verification checking the accuracy between data as recorded in a **source document** and the same data recorded elsewhere (such as a **case record form**)

source document   a document that contains **source data**. This may often be a patient's medical records

source of variation   the reasons why data vary. There are usually many reasons (many sources) including differences between subjects, **interobserver disagreement, intraobserver disagreement** and **measurement error**. Further reasons may be **design effects**, such as **stratification variables**, treatments, etc.

source record $\approx$ source document

Spearman's rank correlation $\approx$ Spearman's rho (ρ)

Spearman's rho (ρ)   a **nonparametric correlation coefficient**. $\Rightarrow$ **Kendall's tau (τ)**

specificity   in a **diagnostic test**, the proportion of subjects who are correctly identified as not having the disease. $\Leftrightarrow$ **sensitivity**

sphericity   the property that all the **variances** and **covariances** in a set of **multivariate data** are equal. This is an important assumption in many forms of analysing **repeated measurements data**. If it does not hold true then the analysis becomes more complex. $\Rightarrow$ **Greenhouse–Geisser correction, Huynh–Feldt correction**

spline function   a function that smoothly joins two **regression lines** which would otherwise have a **break point**

split plot design   this term originated in agricultural **field studies** but, in the context of clinical trials, $\approx$ **crossover design, repeated measurements design**

split plot study   a study designed using a **split plot design**

split-half   half of a data set that has been split into two for the purposes of estimating **reliability**. Note that the term 'half' is not taken literally; the two parts of the data set need not be the same size

split-half reliability   **reliability** of results determined by dividing a data set into two (usually approximately equal halves), determining the appropriate statistical model from one of the two halves of the data (the 'training set') and then checking the consistency of the parameter estimates using the other half of the data (the 'validation set'). $\Leftrightarrow$ **test retest reliability**

Splus®   a computer package commonly used for statistical analysis

sponsor   a person or organisation who gives support (usually financial but not necessarily) to a project. $\Rightarrow$ **pharmaceutical company**

sponsor initiated study   a study initiated by a **sponsor**. In this context, the form of sponsorship is almost always financial and the sponsor is usually a pharmaceutical company. $\Leftrightarrow$ **investigator initiated study**

spontaneous   occurring without being prompted or requested

spontaneous report   a report (usually of an **adverse event**). The report is generally one that comes from clinical practice rather than an adverse event reported as part of a clinical trial

spreadsheet   a simple form of database that may be held on a computer or simply on paper

SPSS®   a computer package commonly used for statistical analysis

spurious   a feature that is observed (usually in a sample of data) but which is not real (is not reflected in the **population**)

square contingency table   a **contingency table** that has the same number of rows as it has columns. The most common is the **two-by-two table**, which has two rows and two columns

square root transformation   the transformation that produces a square root

stability   the property of not changing with time or under changing environmental conditions. This is a desirable aspect of a product. ⇨ **shelf life**

stability data   data to confirm (or otherwise) the **stability** of a product

stability testing   the process of testing a product to determine to what extent it is stable

stable disease   in cancer studies this is generally regarded as a tumour that does not fall into the categories of **complete response, partial response** or **progression**

stacked bar chart   a form of **bar chart** where different subcategories that add to 100% of the main categories are stacked on top of each other (Figure 36). ⇔ **pie chart**

standard deviation   a measure of how spread out a set of data is. It is the square root of the **variance**. ⇔ **standard error**

standard error   a measure of how precisely a parameter has been estimated. It generally increases as the **standard deviation** increases (so as the data become more variable, so the precision of the estimates of any parameters becomes worse); it generally decreases as the sample size increases (so as we have more information, the precision of the estimates of any parameters becomes better). Standard errors are used extensively for constructing **confidence intervals** and **P-values**

standard Normal distribution   a **Normal distribution** with mean equal to zero and **standard deviation** equal to one

standard normal variate   a value from a **standard Normal distribution**

standard operating procedure (SOP)   a written document to describe how various procedures should be carried out in clinical research. Standard operating procedures cover all aspects of clinical trial work from the

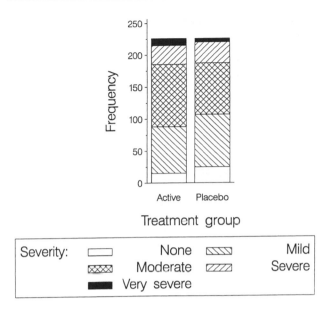

Figure 36 Stacked bar chart. This type of graph can be useful with an ordered categorical scale. It is more useful to estimate the cumulative number of patients with responses in the lower categories than the actual number in any one category. For example, in the Active treatment group, the number of patients with no symptoms is approximately 15; with no or mild symptoms it is approximately 85, etc.

organisation of the company, through protocol writing, running and monitoring studies, data management and statistical analysis, report writing and approval. ⇨ **Good Clinical Practice**

standard score   ≈ *z* **score**

standard treatment   if a treatment is in common use then it is often referred to as 'a standard treatment'. If it is almost universally used then it is sometimes called 'the standard treatment' (note the use of 'a' versus 'the'). ⇨ **gold standard**

standardisation   a broad term. One meaning is simply to transform a set of numbers to *z* **scores**. It is also used for producing various indices, typically of mortality and morbidity

standardised normal deviate   ≈ **standard normal variate**

standardised score   ≈ *z* **score**

STATA®   a computer package commonly used for statistical analysis

statistic   strictly the meaning of this word is 'a function of the data'. The function may be the mean, the maximum value, the fifth smallest value, the difference between the mean and the mode, etc. ⇔ **parameter**

statistical inference   the process of drawing conclusions by using statistical methods

statistical model   see **model**

statistical significance   the claim that is generally made when the calculated *P*-**value** from a statistical **significance test** is less than a prespecified **significance level** (often meaning $P < 0.05$) so that the **null hypothesis** can be rejected

statistical significance test   ≈ **significance test**

statistical test   ≈ **significance test**

statistically significant   the result of a statistical **significance test** when the *P*-**value** is smaller than the prespecified **significance level** and so the **null hypothesis** is **rejected**. ⇔ **clinical significance**

statistics   the art, philosophy and science of using statistical methods to design, manage, analyse and draw conclusions from studies

status   the position of someone or something. When referring to a person, it tends to mean level of seniority (high or low). When referring to a study, it often means the stage that the study has reached: stages might be protocol developed, recruitment started, recruitment part complete, study finished, etc.

StatXact®   a software package that uses 'exact' statistical methods rather than relying on **asymptotic methods**. ⇨ **nonparametric methods**

steady state   stable, or not changing in any systematic way

steering committee   a committee set up to guide (steer) the course of a study, group of studies, organisation, etc. ⇨ **data and safety monitoring committee**

stem and leaf plot   a graphical method for showing a **frequency distribution** (Figure 37). ⇔ **histogram**

stepped wedge design   a design where baseline measurements are repeated several times and different subjects are randomised to treatment after different numbers of baseline observations. It is usually used when a large number of study subjects are all available at the same time and so it staggers the start of randomised treatment across time

stepwise regression   a method of selecting which **predictor variables** should be in a **multiple regression model**. Each variable is included in

```
5    5
5    0 0
4
4    0 0 0 2
3    5 5
3    0 0 0 1 1
2    6 7 8
2    0 0 0 0 0 0 0 0 0 3 3 4
1    5 5 5 6 9
1    0 0 0 0 0 0 1 2 4
0    5 5 5 5 7 8 9
0    1 1 1 1 1 1 1 1 1 2 3 3 4 4 4 4
```

Figure 37 Stem and leaf plot. The number of years each of 66 subjects has suffered from eczema. Reading from the top, one patient has suffered for 55 years, two for 50 years, one for 42 years, three for 40 years, etc.

turn, in a stepwise manner The effect of including and excluding variables is tested. ⇨ *F* **to enter**, *F* **to remove, backward elimination, forward selection**

stepwise selection   ≈ **stepwise regression**

stereogram   ≈ **isometric graph**

stochastic   happening at **random**

stochastic curtailment   stopping recruitment to a study early based on determining the probability that the final result (if the recruitment continued) would not be different from what it is now. ⇨ **premature stopping, sequential study, group sequential study**

stopping boundary   in a **sequential study** or **group sequential study**, the boundary of the **rejection region** for rejecting the **null hypothesis**. Note that, unlike in **fixed sample size designs**, the rejection limit is not a constant value

stopping rule   the rule for deciding when to stop recruitment to a study (usually in the context of **sequential studies** or **group sequential studies**). This may be based on formal statistical considerations or be more informal. ⇨ **alpha spending function, stopping boundary**

stratification   the process of defining and forming **strata**

stratification variable   a **categorical variable** that defines which **stratum** a subject falls in

stratified random sample   a **random sample** that samples separately from different **strata** within the **population**

stratified randomisation   the use of separate **randomisation lists** for different **strata** in the sample. This is often done to ensure that possible **confounding factors** are **balanced** across the treatment groups

stratified sample   ≈ **stratified random sample**

stratified treatment assignment   ≈ **stratified randomisation**

stratify   to divide a sample or population into groups according to the values of a **categorical variable**. This is usually done before **stratified randomisation** or for taking **stratified samples**. ⇨ **poststratification**

stratum   one of the groups formed from stratifying a sample or population

strength of evidence   this is one phrase used to describe the interpretation of a **P-value**. In a statistical **significance test**, the smaller the P-value, the greater the strength of evidence to reject the **null hypothesis**

structured abstract   a requirement by some journals that the abstract (or summary) of a paper should have clear headings rather than being free text. The headings required are usually Introduction, Subjects, Methods, Results, Conclusions. ⇨ **CONSORT**

Student's *t* test   see *t* test

studentised range   when comparing the means of several groups, this is the difference between the largest mean and the smallest mean, divided by the standard error. ⇨ **analysis of variance**

study   a broad term to describe any sort of investigation including clinical trials, surveys, meta-analyses, etc.

study coordinator   the person responsible for organising and coordinating the practical and scientific aspects of a study. It does not usually mean the **principal investigator**, who usually only takes the medical and scientific responsibilities

study data   data that come from a study

study design   ≈ **design**

study group   this does not usually refer to the subjects who take part in a study but to the people who plan, organise, manage, coordinate and run a study

study medication   medication (drugs, etc.) that are being researched in a study. This does not, therefore, include other **concomitant medications**

study nurse   a nurse who is involved in some of the clinical aspects of a study. ⇨ **subinvestigator**

study period   the time during which the clinical aspects of a study are carried out. This is from the time of the first subject being enrolled until the last visit for the last subject

study protocol   ≈ **protocol**

study site coordinator   a **study coordinator** who is responsible for activities at a particular site (or sites) rather than overall

study staff   the staff involved in a study, usually taken as meaning those who work at the study centre and are involved in the clinical aspects of a study rather than those that are involved in the administrative aspects

subcategory   when **categorical data** are categorised by more than one **factor** in a hierarchical way, this produces 'categories within categories'. These are referred to as subcategories. ⇨ **subgroup**

subclass   ≈ **subcategory**

subclinical disease   a disease that does not show any **signs** or **symptoms**. Many diseases may be classed as subclinical in their early stages. ⇨ **sojourn period**

subcutaneous   under the skin. A common route of delivering injections. ⇔ **intramuscular, intravenous**

subgroup   when certain subjects within a randomised group are separately identified (for example, those in certain **strata**), this smaller group is called a subgroup. ⇨ **subcategory**

subgroup analysis   analysis of results of a study just in certain **subgroups**. This may be instead of, or in addition to, an analysis of all subjects. Often subgroup analyses are considered to be potentially misleading

subinvestigator   someone who is qualified to make decisions about subjects in a study but who does so under the supervision of an **investigator**. ⇨ **principal investigator, study nurse**

subject   a person, generally one who takes part in a study. Note that they do not necessarily have to have any disease (⇔ **patient**) and they may or may not be taking part voluntarily (⇔ **volunteer**)

subject id   a unique reference code given to a subject in a study. It may be numeric or alphanumeric

subject identification number   a **subject id** that is numeric

subjective   general impressions rather than confirmed facts. ⇔ **objective**

subjective Bayes   **Bayesian** statistical methods where the **prior distribution** is formed from **subjective data**. ⇔ **empirical Bayes**

subjective data   values that are formed from impressions rather than from precise measurements. Examples include data recorded on **visual analogue scales**, data recorded in the form of 'mild', 'moderate' or 'severe' symptoms and data on **likert scales**. ⇔ **objective data**

subjective endpoint   an endpoint to a study that is **subjective data**. ⇔ **objective endpoint**

subjective measurement   a measurement of **subjective data**. ⇔ **objective measurement**

subjective outcome   an **outcome** that is **subjective data**. ⇔ **objective outcome**

subjective probability   this usually refers to a **prior probability** in **Bayesian** statistics. ⇨ **subjective Bayes**

sublingual   under the tongue

subscript   in typesetting, a character that is set below the line of other characters, for example the 'i' in x_i. ⇔ **superscript**. Subscripts and superscripts are used extensively in mathematical and statistical notation and in chemical formulae

subset   ≈ **subgroup**

sum of squares   in a set of data, the sum of the squared deviations of each value from the mean. Sums of squares are used extensively in **analysis of variance** and **regression**. ⇨ **least squares**

summary measure   a **statistic** (a function of the data) used to summarise **repeated measurements**. Examples include **area under the curve**, C_{max}, T_{max}

summary of product characteristics (SPC)   all of the licensing information about a product: prescribing information, contraindications, etc. ⇨ **package insert**

summary result   ≈ **statistic**

summary statistic   ≈ **statistic**

supercomputer   a very powerful mainframe computer. ⇨ **microcomputer**, **minicomputer**, **mainframe computer**

superinfection   in studies of antibiotics, a superinfection is a new infection that appears in addition to a pre-existing one

superiority study   a study where the objective is to show that one treatment is better than another. In general the **null hypothesis** would be stated as $H_0 : \mu_1 = \mu_2$ and the **alternative hypothesis** would be $H_1 : \mu_1 > \mu_2$. ⇨ **equivalence study**, **noninferiority study**

superposition   for drugs with **linear kinetics**, the concentration of the drug at any time following multiple doses is the sum of the concentrations that would be observed had each of the doses been given as a single dose

superscript   in typesetting, a character that is set above the line of other characters, for example the 'y' in x^y. ⇔ **subscript**. Superscripts and subscripts are used extensively in mathematical and statistical notation and chemical formulae

supine   lying down on one's back. ⇔ **prone**

supporting analysis   a **secondary analysis** that gives similar information to the **primary analysis** and so helps justify (or support) conclusions based on the primary analysis

supporting study   a study that is not definitive but helps to support the results and conclusions of a **confirmatory study**

suppository   medication delivered by insertion into the vagina or rectum. ⇨ **pessary**

surgery   the branch of medicine that treats diseases by operative procedures

surgical   methods that use surgery

surgical study   a study that compares different surgical procedures or compares surgical and nonsurgical (possibly medical) procedures

surgical treatment   treating a patient by surgery. As with **placebo treatment**, this illustrates the very broad use of the term 'treatment'

surrogate   a substitute. A variable that is a surrogate for the most clinically meaningful **endpoint**. In hypertension studies, the most important endpoint would usually be **mortality** (possibly restricted to cardiovascular reasons) but raised blood pressure would often be the endpoint that is measured. Blood pressure is being used as a surrogate for mortality

surrogate endpoint   an **endpoint** that is a **surrogate**

surrogate marker   ≈ **surrogate**

surrogate observation   an observation that is a **surrogate**

surrogate outcome   an outcome that is a **surrogate**

surrogate variable   ≈ **surrogate**

surveillance   the act of watching over; most commonly used in the context of **post marketing surveillance study**

survival   remaining alive. If death is not the **outcome** of a study then 'survival' is sometimes broadened to include whatever the outcome is defined to be

survival analysis   methods of analysis of **survival studies** that use the time to death (or time to some other event) as the **response variable**. Methods include **Cox's proportional hazards model**, the **proportional odds model**, the **log rank test**

survival curve   a graph showing the proportion of subjects in a **survival study** who are still alive as time progresses (Figure 38)

survival data   data relating to the length of time subjects 'survive' in a study. ⇨ **survival time**

survival function   the equation that describes the **survival curve**

survival rate   the proportion of subjects in a study who are still alive at a particular time. Note that the term 'death rate', although seemingly the opposite of survival rate, is less often used. ⇨ **hazard rate**

survival study   a study of the time until an event (often, but not restricted to, death) occurs

survival time   the length of time a subject survives after some intervention

suspension   a form of **presentation** of product in liquid form, for oral

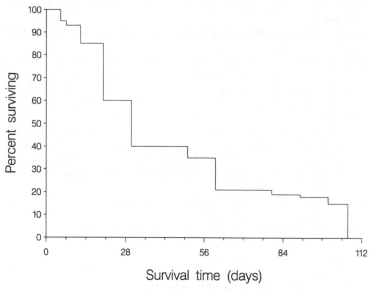

Figure 38   Survival curve. This graph simply shows the percentage of patients surviving in a study of pancreatic cancer

consumption. ⇨ other **delivery devices** such as **capsule, tablet, transdermal patch**

symmetric distribution   a distribution (usually a **probability distribution** rather than a **frequency distribution**) that is symmetric. This means that the mean, the median and the mode will all be equal. The **Normal distribution, *t* distribution** and **uniform distribution** are all symmetric. It is sometimes wrongly thought that a symmetric distribution must be a Normal distribution but this is clearly not the case

symmetric test   ≈ **two sided test**

symptom   a feature of a disease or illness that a patient experiences (headache, nausea, etc.). ⇔ **sign**

symptom checklist   a list of possible symptoms that a subject is asked to review and identify which (if any) they are suffering from

syndrome   a group of signs and symptoms that, when occurring together, characterise a disease. ⇨ **diagnosis**

synergistic effect   the effect of two (or more) interventions that is greater than the sum of the individual effects. ⇨ **interaction**

synergy   more than one intervention acting together (usually in a positive way). ⇔ **drug interaction**

synthesise   to draw information from different sources together and make judgements. ⇨ **overview, meta-analysis**

syringe   a **delivery device** for injecting liquid into the body. ⇨ **intravenous, intramuscular, subcutaneous**

system organ class (SOC)   a classification of **signs**, **symptoms** and diseases according to different organ systems of the body (cardiovascular, central nervous system, endocrine, etc.). ⇨ **high level term, preferred term**

systematic   occurring in a regular, steady, manner

systematic allocation   a form of treatment allocation that is not random but follows a regular pattern. For example, all subjects recruited on days of the month that are odd (1st, 3rd, 5th, etc.) may be given one treatment whilst all subjects recruited on the even numbered days are given the alternative treatment. ⇨ **systematic sample**

systematic assignment   ≈ **systematic allocation**

systematic error   an error that is always made. Examples include measuring subjects' weight on a faulty pair of scales. Often synonymous with **bias**

systematic review   a thorough and complete review and assessment of all the literature on a subject. The term is often used in the context of **meta-analyses**

systematic sample   a sample from a population that is not taken at random. A simple method for taking a systematic sample would be to select every tenth patient that enters a doctor's surgery. ⇨ **systematic allocation**

systematic variation   variation that is the same (or very similar) between all subjects in a study or that is inherent in a measuring instrument. ⇔ **random variation**

systemic   relating to the entire body rather than individual parts of it. ⇔ **topical**

***t* distribution**   the **probability distribution** for the *t* **statistic**

***t* statistic**   the calculated value of *t* (as in *t* **test**) from a set of data

***t* test**   a statistical **significance test** about the mean of a population when the variance has to be estimated from the data. ⇨ **independent samples *t* test, paired *t* test, single sample *t* test**

table   a summary of results presented in rows and columns

table shell   ≈ **ghost table**

tablet   a dissolvable pill made of compressed powder that contains a drug. ⇨ other **delivery devices** such as **capsule, transdermal patch**

tabular   in the form of a **table**

tabulate   to produce a **table** of data or of results

tabulation   ≈ **table**

tachyphylaxis   lessening of the effect of a treatment after prolonged use, or the need to increase the dose in order to maintain a given effect

tail   the **extreme values** (highest and lowest) of a **distribution**. The term is only generally used when the frequency (or **relative frequency**) becomes smaller and smaller as the values become more extreme; so it is not usually applied to **U shaped distributions**, for example

tail area   the area under a **probability distribution** curve in the **tail** of the distribution beyond a specified **cutoff point**. ⇨ **rejection region**

tail area probability   the **probability** that a **sample statistic** will lie in the **tail area** of a **probability distribution**. ⇨ *P*-value, **significance level**

target population   the **population** from which it is intended to **sample** or to which conclusions from a study are intended to apply

telephone interview   an **interview** carried out over the telephone

telephone randomisation   the process of **central randomisation** is often carried out by the investigator contacting a central office by telephone. This is referred to as telephone randomisation. At the central office there may be a computer that answers calls and gets information via a touch tone telephone, or there may be an operator who takes calls and deals with them manually. Both cases fall under the heading of telephone randomisation

temporal   happening with time

temporal variation   variation that occurs with time. ⇨ **secular**

teratogenic   a drug (or other procedure) that causes harm to a foetus

tercile   each of the $33\frac{1}{3}$rd and $66\frac{2}{3}$rd **centiles**

term   a word or expression that has a specific meaning in a particular discipline. Also a single part of a mathematical model

terminal   life ending. Also used to mean a device for inputting to, and getting output from, a computer

tertiary   of third importance (behind **primary** and **secondary**). Often secondary is used to encompass everything that is not primary; if tertiary is used, it is often used to encompass everything that is not primary or secondary

tertiary care centre   a clinic or other centre where patients are treated or cared for after they have been referred from a **secondary care centre** (and so also after being referred from a **primary care centre)**

tertiary care study   a study carried out in a **tertiary care centre**

tertile   ≈ **tercile**

test   a way of challenging an idea to see if it is true. This applies to medical tests (such as **diagnostic tests**) and to statistical tests (as in **hypothesis test** and **significance tests**). The term is also used more generally simply to find an answer to a simple question such as what is someone's blood pressure

test case   any single member of the **test group**. ⇨ **case**

test group   a group of subjects who are being tested or experimented on. In the case of a study of two treatments, the term is usually used to refer to the group who receive the **experimental treatment**. ⇔ **control group**

test of hypothesis   ≈ **hypothesis test, significance test**

test of significance   ≈ **significance test, hypothesis test**

test result   the result of a **test**. Usually this term is restricted to results of medical tests (**diagnostic tests**, laboratory tests, etc.) rather than being applied to results of statistical **significance tests**

test retest reliability   the **correlation** between two measurements on the same subject. ⇨ **within subjects variance**

test run   a test of practical issues with carrying out a study. ⇨ **pilot study**

test statistic   the immediate result of carrying out a statistical **significance test**. The test statistic is then compared with the appropriate **probability distribution** to see how extreme it is (to what degree it lies in the **tails** of the distribution). From here, the *P*-value can be calculated

test treatment   ≈ **experimental treatment**

test validity   the extent to which a test measures what it is supposed to be measuring. This is particularly important in questionnaires such as those that measure **quality of life**, where the 'real' answer is almost impossible to know

theorem   a statement or conclusion in mathematics deduced from other formulae. Note that theorems are generally considered to be true. ⇔ **theory**

theory   a supposition or idea. ⇨ **hypothesis**. ⇔ **theorem**

therapeutic   concerned with treating a disease

therapeutic dose   the dose of a drug that is effective for treating a disease. ⇨ **therapeutic range**

therapeutic effect   an **effect** caused by a therapy (or treatment). Usually this is considered to be the beneficial, not the adverse, effects

therapeutic equivalence   the same **therapeutic effect**. Two products may differ, may require different doses and **formulations** but may still be therapeutically equivalent

therapeutic index   ≈ **therapeutic ratio**

therapeutic range   the range of doses of a drug that give a **therapeutic effect**

therapeutic ratio   the ratio of **beneficial effect** to **adverse events**. ⇨ **risk–benefit ratio**

therapeutic study   a study of therapies. Used in place of **clinical trial** when it is required to emphasise that the study is one of treatment of a disease rather than of prevention or **prophylaxis**

therapeutic window   ≈ **therapeutic range**

therapy   ≈ **treatment**

three factor interaction   the **interaction** between three distinct **factors**. These might be, for example, treatment, gender and stage of disease. Such interactions are rare, difficult to detect with any certainty and often difficult to justify on clinical grounds. **Two factor interactions** are also rare, but less so. ⇨ **high order interaction, factorial study**

three period crossover   a **crossover study** that has three **treatment periods**. This may include a comparison of three treatments or of only two. In the case of comparing two treatments, subjects would be given the same treatment in two of the periods and a different treatment in the third

three way interaction   ≈ **three factor interaction**

threshold   ≈ **cutoff point**

tied observations   two data values that are equal. The term is usually restricted to use with **ordered categorical data**

time dependent covariate   a **covariate** that changes with time. Usually it is assumed that covariates are fixed: age at entry to a study might be a covariate that, once measured, does not change. Exposure to possible carcinogens in the workplace is a covariate that might change with time as an employee works at different sites and as processes at those sites change. ⇔ **time independent covariate**

time independent covariate   a **covariate** that does not changes with time, such as gestational age at birth, sex, ethnic group, baseline disease severity. Most use of the word covariate assumes time independent covariate. ⇔ **time dependent covariate**

time interval   a specific, identifiable, period of time. ⇔ **time point**

time period   ≈ **time interval**

time point   a specific, identifiable, moment in time. The definition of 'moment' may vary with the context: in some cases it might be a scheduled study visit, in others it might be a particular day or the instant at which an injection is administered. ⇔ **time interval**

time series   a series of data that are collected **sequentially** in time. Usually it is assumed that a time series is very 'long' in the sense that it contains hundreds of sequential measurements. Contrast this with **longitudinal data**, where generally fewer time points are assumed. **Repeated measurements** data might have the same interpretation as longitudinal data or may involve even fewer **time points**

time window   ≈ **time interval**

titrate   to search for the **optimum** dose of drug in the body to produce a required **therapeutic effect**

titration   the process of **titrating**

titration period   a period during a study when each subject is being **titrated**. A consequence of studies that have titration periods is that each subject can potentially end up on a different dose of drug. If this period is then followed by a **treatment period**, the results of the study will not be in terms of an effect produced by a given dose of the drug but rather an effect given by the optimum dose of the drug, as defined for each subject separately

titration study   ≈ **dose escalation study**

titre   the **concentration** of a drug

T_{max} the time at which the maximum **concentration** of a drug occurs in the body (see Figure 1, **area under the curve**). ⇨ C_{max}

tolerability   ≈ **safety data**

tolerance   the allowable (and acceptable) difference between a measured value and the **true value**. ⇨ **accuracy**. The term is also used to mean acceptability (of a product)

tolerance limit   the maximum acceptable difference between a measurement and its **true value**. ⇨ **accuracy**

topical   local; relating to specific parts of the body rather than to all of it. **Ointments** and **creams**, for example, are topical treatments. ⇔ **systemic**

total cost   the sum of all costs from all sources. ⇨ **direct cost, indirect cost**

total sum of squares   the **sum of squares** of a set of data, ignoring any strata (treatment groups or other classification variables). ⇨ **between groups sum of squares, within groups sum of squares, between subjects sum of squares, within subjects sum of squares**

toxic   poisonous

toxicity   the extent to which a substance is **toxic**

toxic effect   a poisonous effect. ⇨ **adverse reaction**

toxicology   the study of poisons or of poisonous effects of drugs

toxin   a poison

trace   the smallest amount of a chemical (drug or otherwise) that can be detected. It is usually so small that the actual amount cannot be determined; only its presence can be detected

trade name   the name given to a product for the purpose of marketing. ⇨ **generic name**

training set   see **split-half reliability**

transcribe   to copy from one place to another. Common examples are the copying of details from patients' notes into a **case record form** or copying data on case record forms into a computer

transcription error   an error that is made when **transcribing** data

transdermal patch   a method of delivering drug to the body. The drug is impregnated into a patch that is placed on the skin and the drug is absorbed through the skin

transform   to change, usually in the sense of changing numbers through mathematical functions

transformation   the result of transforming. ⇨ **function**

transition matrix   ≈ **shift table**

treat   to give a medical treatment to someone with a disease. In the context of clinical trials, the term is often used broadly to include surgical interventions, giving subjects **placebo** and giving medication to subjects who are not **patients**

treatment   any form of **intervention** given to **patients**. The term is used extremely widely in the context of clinical trials. It includes use of **placebos, palliative care**, even the use of 'no treatment' if that is one of the **treatment regimens** being evaluated. The term is also sometimes used when referring to subjects who are not ill and who are not, therefore, patients

treatment allocation   the treatment that has been allocated to a subject (by **randomisation** or some other means)

treatment allocation ratio   the ratio of how many subjects are assigned (by **randomisation** or by other means) to each **treatment group** in a study. The most common situation is that each treatment group has the same number of subjects (1:1 allocation, or **equal allocation**) but 2:1 and 3:1 allocations are sometimes used

treatment arm   ≈ **treatment group**

treatment comparison   the **comparison** between two (or more) **treatment groups**, based on one of the **endpoints** of a study

treatment difference   the difference between two **treatment groups** based on one of the study **endpoints**. This might commonly be the difference in mean response, difference in the proportion of responders, difference in median **survival times**, etc. ⇨ **effect, treatment effect**

treatment effect   largely synonymous with **treatment difference** but estimates such as **odds ratios** tend to be referred to as treatment effects rather than as treatment differences

treatment emergent signs and symptoms (TESS)   a method of defining an **adverse event**. If a subject has any **signs** or **symptoms** of a disease before entering a study, continuation of those signs or symptoms is not recorded as an adverse event. Worsening of any of those signs or symptoms is regarded as an adverse event, as is the emergence of any new sign or symptom

treatment failure   a subject who has a poor **response** to treatment (generally a response that has been declared to be not sufficiently good). ⇔ **treatment success**

treatment group   the group of subjects that are assigned to receive one particular **treatment** (treatment is used here in the very broadest sense). The term is also sometimes used to refer specifically to those subjects who do not receive placebo. In this case, the term **placebo group** is often used to describe subjects who receive placebo

treatment group comparison   ≈ **treatment comparison**

treatment interaction   an **interaction** between two (or more) **treatments**

treatment interaction effect   the size of a **treatment interaction**

treatment period   the **time period** during which **treatment** is given. ⇔ **run in period, washout period, follow-up period**

treatment regimen   a term to encompass **treatment, dose**, method of delivery, time of delivery, etc. For example a treatment regimen might consist of: 2 weeks of tablets, three times a day; followed by 10 days of injections once a day; followed by 6 weeks of physiotherapy

treatment schedule   a time schedule for when subjects are supposed to receive **treatment**

treatment sequence   the order in which **treatments** are given (multiple treatments either to the same subject, as in a **crossover study**, or to different subjects in a **parallel group study**)

treatment success   a subject who has a good response to treatment. ⟺ **treatment failure**

treatment–period interaction   an **interaction** between **treatment** and the **period** when it is given. This implies that the treatment effect is different during different time periods in the study. This is particularly important in **crossover studies**

trend   a general course. The term is used variously: it is used of **longitudinal data** to describe a consistent change with time; it may be used in studies with multiple doses of the same drug to imply a consistent change in **effect** with increasing (or decreasing **dose**); sometimes it is used in simple comparisons between two groups when a difference is observed but that difference has not been shown to be **statistically significant** ('there was a trend towards statistical significance'). In this last situation, it is used badly and tends to reflect the investigator's hopes, rather than what the data have actually demonstrated

triage   classification of patients according to their need and their likely **prognosis**. Briefly, those with minor injuries can wait; those with very severe injuries are likely to be past helping; those in the middle category are the ones who should be treated first

trial   ≈ **clinical trial**

trial and error   the informal process of trying one idea to see if it works; if it does not then trying another idea and so on until a solution to a problem is found

trialist   a person who has experience and expertise in scientific and practical aspects of running clinical trials

triangular test   a **closed sequential study** design where the **baseline** (zero recruitment) and **stopping boundaries** form a triangle

trim   to delete **outliers**. ⟺ **truncate**

trimmed mean   the mean of a sample of data after **outliers** have been deleted. A (small) percentage of the data values are deleted from each **tail** of the **frequency distribution**. ⟹ **truncated mean, Winsorised mean**

triple blind   the situation where the subject is blinded to the study medication, the investigator is blinded and the data management staff are blinded. ⟹ **single blind, double blind, quadruple blind**

triple mask   ≈ **triple blind**

trough   on a graph, the opposite of a **peak**

trough value   the minimum value from a set of selected data. ⇔ **peak value**

trouser leg design   an informal term for some types of **sequential design** where the stopping boundaries appear, graphically, like the outline of a pair of trousers (≈ **closed sequential design**, Figure 4)

true mean   the mean of the **population**, rather than that of the **sample**. The **parameter** value rather than the sample **estimate**. The true mean is generally never known although in the case of **finite** sized populations it could, in theory, be determined. ⇔ **sample mean**

true value   see, for example, **true mean, true variance**. ⇔ **statistic**

true variance   as with **true mean**, this is the **population** value of the variance. ⇔ **sample variance**

truncate   to delete **extreme values** from a **distribution**. The values deleted are not necessarily **outliers**. ⇔ **trim**

truncated data   data from which the **extreme values** have been deleted

truncated distribution   a **distribution** from which the **extreme values** have been deleted. This can apply to both **frequency distributions** and **probability distributions**

truncated mean   the mean of a set of truncated data. ⇨ **trimmed mean, Winsorised mean**

truncated observation   ≈ **truncated data**

Tukey's least significance difference (LSD) test   a **multiple comparison method** for comparing the means of several groups

two armed study   a study with two treatment groups, usually assumed to be a **parallel groups study**

two compartment model   a theoretical model used for describing the way that many drugs are dispersed around the body. It assumes that drug is absorbed into one compartment of the body (typically into **plasma**) and then into other organs and tissues of the body. ⇨ **first pass metabolism**

two factor interaction   the **interaction** between only two **factors** (such as gender and treatment). ⇨ **three factor interaction, high order interaction, factorial study**

two period crossover   ≈ **two period crossover design**

two period crossover design   a **crossover study** with two **treatment periods**, generally to compare two treatments

two period crossover study   a **crossover study** designed as a **two period crossover design**

two sample *t* test   ≈ **independent samples *t* test**

two sided   concerned with both **tails** of a **distribution**. ⇔ **one sided**

two sided alternative   the **alternative hypothesis** that is a **two sided hypothesis**. ⇔ **one sided alternative**

two sided hypothesis   an **alternative hypothesis** for the comparison of the **means** (or other **parameter**) of two or more groups. The **null hypothesis** is usually that the two group means are equal; the **alternative hypothesis** is that the two group means are not equal. This allows either group to be declared superior. ⇔ **one sided hypothesis**

two sided test   any statistical **significance test** where the **alternative hypothesis** is a **two sided hypothesis**. ⇔ **one sided test**

two stage stopping rule   a **stopping rule** that has two criteria that both need to be fulfilled in order to stop

two tailed   ≈ **two sided**

two tailed alternative   ≈ **two sided alternative**

two tailed hypothesis   ≈ **two sided hypothesis**

two tailed test   ≈ **two sided test**

two way analysis of variance   analysis of the mean of a **continuous variable** by a model that has two independent **categorical variables**. ⇨ **analysis of covariance, one way analysis of variance, regression**

two way classification   cross-classification of data by two **categorical variables**. For example, the mean response might be classified by treatment group and by gender

two way interaction   ≈ **two factor interaction**

two way table   a **table** (of **counts, means** or any other **statistic**) that has two **classification variables**

two-by-two contingency table   ≈ **two-by-two table**

two-by-two table   a **table** of data or summary statistics with two rows and two columns. ⇨ **odds ratio, risk ratio**

Type I error   in a statistical **significance test**, the probability that the **null hypothesis** is **rejected** when it is actually true. ⇔ **Type II error**. ⇨ **Type III error**

Type II error   in a statistical **significance test**, the probability that the **null hypothesis** is not **rejected** when it is actually false. This often happens due to lack of **power**. ⇔ **Type I error**. ⇨ **Type III error**

Type III error   in a statistical **significance test** the probability that the **null hypothesis** is **rejected** and Treatment A is declared to be superior to Treatment B but the true state is that Treatment B is superior to Treatment A. ⇨ **Type I error, Type II error**. Some people also use the term to refer to the correct answer to the wrong question

U shaped curve   a curve having two **peaks** and a **trough** in the middle (Figure 39)

U shaped distribution   a **distribution** that has **peaks** at (or very near) the minimum and maximum values and a low **relative frequency** (a **trough**) in the middle. Such distributions are not common but one example is the distribution of alcohol intake in some populations

unadjusted   **parameter estimates** that are not **adjusted** for any **covariates** or **confounders**

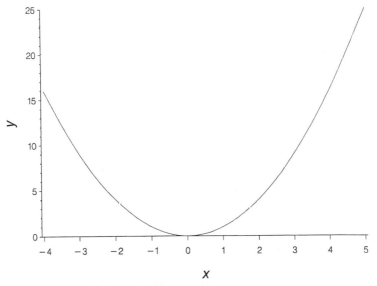

Figure 39   U shaped curve. Any curve of this general shape is called U shaped. This example is also a quadratic curve

unbalanced　a study (or sometimes just one aspect of a study) that is not **balanced** (either intentionally or inadvertently)

unbalanced block design　a study design that has used **blocking** but which has not used **balanced blocks**

unbalanced design　a study design that is not **balanced** (either inadvertently or intentionally) as with, for example, a **partially balanced design**

unbalanced treatment design　≈ **unbalanced design**

unbiased　not **biased**

unbiased estimator　an **estimator** that is not **biased**

unblind　to open the **randomisation schedule** in a blinded study; also a term that describes a study that is **open label**

uncontrolled　not **controlled**; having no **control group**

uncontrolled study　a study that has no **control group**

under represent　not represent fully. This particularly applies to the **sample** of subjects in a study and how well the distribution of **demographic data** reflects the demography of the **population**. ⇨ **sample demographic fraction, population demographic fraction**

underlying distribution　an informal term for the **probability distribution** that can never be properly observed; only a sample forming a **frequency distribution** can be observed

unequal allocation　allocation to **treatment groups** that is not in an equal proportion in each group. ⇨ **allocation ratio**. ⇔ **equal allocation**

unequal randomisation　the same as **unequal allocation** but this stresses that the allocation method was one of **randomisation**

unexpected adverse drug reaction　an **adverse reaction** that is not expected and does not appear in the **investigator's brochure** or **summary of product characteristics**. ⇨ **unexpected adverse event**

unexpected adverse event　an **adverse event** that is not expected and does not appear in the **investigator's brochure** or **summary of product characteristics**. ⇨ **unexpected adverse drug reaction**

uniform distribution　a **probability distribution** where all values from the distribution have equal **probability** of occurring (Figure 40)

uniform prior　in **Bayesian** statistics a **prior distribution** that assigns equal **probability** to all values of a **parameter** between two limits. Often (but not necessarily) the limits may be from **minus infinity** to **plus infinity**. ⇨ **reference prior**

uniform treatment assignment　≈ **equal allocation**

uniformly most powerful test　any statistical **significance test** that has as much **power** as any other test for testing the same **hypothesis**. Note that,

0   1

Figure 40   Uniform distribution. A very simple probability distribution. All values between 0 and 1 have equal probability of being observed. (The range need not necessarily lie between 0 and 1.) This type of distribution would rarely be seen in a sample of data but does have some useful theoretical applications

in general, **two sided tests** never have a single uniformly most powerful test but **one sided tests** may have

unimodal   having one **mode**

unimodal distribution   a **distribution** that has only one **mode**. ⇔ **bimodal**

uninformative prior   ≈ **reference prior**

unique   the only one of a kind

unit normal variable   ≈ **standard normal variate**

unit of analysis   in most situations, each subject in a study will be analysed as an **independent** subject or unit of analysis. In some situations where each subject is not **randomised** individually (see, for example, **cluster randomisation**) then the unit of analysis needs to reflect this. ⇨ **experimental unit**

univariate relating to only one variable (usually meaning one **response variable**). ⇔ **bivariate, multivariate**

univariate analysis methods of analysis that assume a single **response variable**. ⇔ **bivariate analysis, multivariate analysis**

univariate data one **response variable**. ⇔ **bivariate data, multivariate data**

univariate distribution the **distribution** of one variable. ⇔ **bivariate distribution, multivariate distribution**

universe the totality. This term is sometimes used synonymously with **population** and sometimes regarding all possible values of a **parameter** or all possible **models** that might be considered

unpaired *t* test ≈ **independent samples *t* test**

unrestricted randomisation **randomisation** (to **treatment groups**) that has no restrictions (like **blocks** or **strata**). ⇔ **restricted randomisation**. ⇨ **completely randomised design**

up and down design a study design that can increase or decrease the dose of medication until an optimal dose is found. ⇨ **dose escalation study**

update to take new information and use it to bring old decisions, judgements, etc. up to date. The term is often used in **Bayesian** statistics to describe updating a **prior distribution** to obtain a **posterior distribution**

upload copying datafiles, programs, etc. from a local computer to a central computer. ⇔ **download**. ⇨ **distributed data entry**

upper quartile the 75th **centile**. ⇨ **lower quartile, median**

urinalysis measurement and analysis of chemicals in urine. ⇨ **laboratory data**

urn a pot (usually in a hypothetical sense) that is considered to contain variously coloured balls and is used for experiments about **sampling**

urn model a statistical model which assumes that the underlying process is that similar to sampling different coloured balls from an **urn**. The different colours may represent different treatments of different events that might occur

urn model randomisation a **randomisation** method that behaves in the same way as an **urn model**

utility an overall assessment of the value of something gained by amalgamating all forms of **benefits** and **costs**

utility analysis the analysis of **utilities**

utility function a mathematical function that summarises **utility**

vaccine extracts of (usually) dead bacteriological cells used as a **preventive treatment** against some kinds of **infections**

vaccine study a study of the **efficacy** or **safety** of a **vaccine**

vague prior in **Bayesian** statistics a **prior distribution** that shows great uncertainty in the value of a **parameter**. Not as extreme as a **reference prior**

valid reasonable or logical (and assumed to be correct) ⇨ **validate**

valid analysis an analysis of data that is **valid**. Note, however, that whilst it is assumed that valid analyses give the correct answers, this is not necessarily so

valid cases analysis ≈ **per protocol population**

validate to confirm that something (a measuring device, scale, machine, computer program etc.) is giving correct answers. This may often be done by comparing the results with those from a **gold standard** of measurement. ⇨ **valid**. The term is also used in the sense of confirming that data are correct although, in this case, the extent of validation is often restricted to ensuring the data are plausible

validation the process of **validating**

validation set see **split-half reliability**

validity the extent to which a process or measurement is **valid**

validity check ≈ **edit check**

value a single item of data. Synonymous with terms such as data value and data item

variability the situation that almost always arises when the same data, measured at different times, on different people or by different methods, yield different answers. This variation is often described **quantitatively** by the **range**, the **interquartile range** or the **standard deviation**

variable the mathematical term for a **characteristic** or property of something or someone that is being measured. It may vary from time to time and from subject to subject. ⇨ **random variable**

variable cost ≈ **per unit cost**

variable selection method any of the methods of deciding which variables (sometimes in this context referred to as **covariates**) should be included in a **multiple regression model**. Specific methods include **backward elimination, forward selection, stepwise regression**

variance   a measure of the **variability** in **numeric** data. It is the square of the **standard deviation**

variance component   when the variability in data is caused by several **factors** (for example, variability between treatments, between subjects, between times, **random variation**), the proportion of variation due to each of these factors can be determined and each of these separate amounts is referred to as a variance component. ⇨ **components of variance**

variance ratio   the ratio of two **variances**. This is commonly used in **analysis of variance**

variance ratio test   ≈ *F* **test**

variance stabilising transformation   any **transformation** of data that is used to keep the **variance** of each observation approximately equal to that of all the other observations. The most commonly used are logarithms and the square root. ⇨ **heteroscedastic**

variance–covariance matrix   in **multivariate data**, a square **matrix** whose elements are the **variances** of all the variables and the **covariances** between all pairs of variables

variant   showing **variation**

variate   a **variable**

variation   ≈ **variability**

vector   a column (column vector as in Table 15a) or row (row vector as in Table 15b) of numbers, **statistics**, **parameters**, etc. Each of the rows or columns of a **matrix**

vehicle   a term frequently used in dermatology studies to describe the **cream, ointment**, etc. that contains the active ingredients of a drug. The cream is necessary to help apply the active ingredients and so is sometimes called a vehicle. ⇨ **delivery device, placebo**

verbal assent   ≈ **oral assent**

verbal consent   ≈ **oral consent**

verification   the process of **verifying**

verify   to confirm that something is correct, either by a simple double check or by using a different method of measurement

vial   a small container (usually glass) that contains liquid medication

violate   to break rules (often used with reference to **inclusion criteria** and **exclusion criteria**)

violation   an instance of where rules (**inclusion criteria, exclusion criteria**, for example) have been broken

Table 15   Column vector (Table 15a) of heart rates (beats/minute) for six healthy subjects and row vector (Table 15b) of heart rates for subject number one on four occasions

(a) Column vector	(b) Row vector
$\begin{bmatrix} 78 \\ 67 \\ 75 \\ 80 \\ 56 \\ 72 \end{bmatrix}$	$[78,75,78,72]$

virus   small micro-organisms that enter cells and cause disease. ⇨ **bacterium**. In computer terms, a program that is intended to maliciously interfere with the running of the computer. Some of these viruses cause little damage but are very inconvenient; others may cause complete loss of data, **software** and even damage to **hardware**

visit   the term used to refer to each occasion when a subject in a study meets with the **investigator** (or other **study staff** such as a **study nurse**). This usually applies only to **outpatient studies**. ⇨ **baseline visit, randomisation visit, end of treatment visit, follow-up visit**

visit data   the data relating to a single **visit** in a study

visit date   the date on which a **visit** is due to occur or actually does occur

visual analogue scale   a way of measuring **subjective** feelings (such as happiness, pain, fatigue, etc.) by asking subjects (sometimes also investigators) to mark on a line how good or bad they feel about the variable in question. Traditionally the line is 10 cm long but this is arbitrary. Each end of the line is described in terms such as 'the best possible . . .' and 'the worst possible . . .' (Figure 41)

Visual Basic   a high level computer programming language. ⇨ **BASIC, C, C++, Fortran**

vital signs   the most important **signs** relating to life. These are usually considered to be blood pressure, body temperature, pulse and respiratory rate

vital statistics   **demographic data** of a **population** commonly relating to birth, death, marriage, divorce and health status

volume of distribution   the ratio of the volume of drug in the body to its **concentration** in the blood stream

volunteer   someone who does something through their own choice. Note that under most circumstances all subjects who take part in trials

This treatment is
the best I have
ever had

This treatment is
the worst I have
ever had

Figure 41   Visual analogue scale. A simple scale, in this example for enquiring about patients' opinions of a treatment

should do so on a voluntary basis, and it is not helpful to use this term to describe those who take part in studies. Healthy subjects in studies may (or may not) be volunteers; patients in studies may (or may not) be volunteers

volunteer bias   a **bias** that is often caused because subjects who **volunteer** for studies may not be representative of the **population** to which the results of the study are intended to apply. This affects **external consistency** more than **internal consistency**. ⇨ **healthy worker effect, sample demographic fraction, population demographic fraction**

vulnerable subject   a subject who is at risk (from **trialists** rather than from disease). Examples include patients with very serious diseases and with high expectation of the benefits of a new product, subjects who have a working relationship with the investigators (medical students, nurses, employees of pharmaceutical companies, etc.)

washout   the process of allowing time for drugs to be naturally **excreted** from the body. This is often necessary (or desirable) at the beginning of a study to ensure that the effects of any medication the subject may have been taking before enrolling in the study do not interfere with the assessment of the study drugs

washout period   the **time interval** during which **washout** is allowed to take place

Weber–Fecher law   a law that states that the **absolute change** in the effect of a drug is proportional to the **relative change** in dose

Weibull distribution   a **probability distribution** commonly used as the assumed distribution of **survival times**

weight   a term used when different data points are given different amounts of importance. Weights may sometimes be chosen as being inversely proportional to the **variance**: this implies that data values with a small variance (that is, those which are very **accurate**) are given more importance in calculating **parameter estimates** than those with higher variances (or less accuracy)

weight of evidence $\approx$ **strength of evidence**

weighted average   an average ($\approx$ **mean**) that has been calculated using different **weights** for different data values

weighted least squares   analyses using the general methods of **least squares** but where the calculations allow different **weights** for different data values

weighted mean $\approx$ **weighted average**

weighting   the process of applying **weights** to data

well controlled   said of a study that has very thorough design constraints and which is carried out exactly as specified in the **protocol**. The term generally does not refer specifically to the use of **control groups** although that would be one of the features of good design

wellbeing   a person's general state of health (but not confined to good health)

Wennberg's design   a study design where subjects are randomly assigned to a group who receive the treatment of their own choice or to a group who are then randomly allocated to one or other study treatments. ⇨ **Zelen's randomised consent design**

what you see is what you get   a computer term (sometimes also used elsewhere) to indicate that the output shown on a screen is exactly as will appear on a printed version

whistle blower   a derogatory term for someone who reports a suspected case of **fraud** or misconduct

white noise   **random**, unexplainable, **variation**, usually assumed to be from a **Normal distribution** with mean zero. ⇨ **variability**

WHO-ART   a coding system of **adverse events**. Stands for World Health Organization Adverse Reaction Terminology. ⇨ **COSTART, MedDRA**

wide area network   similar to **local area network** but covering a wider geographical area

Wilcoxon matched pairs signed rank test   a **nonparametric** statistical **significance test** for testing the **null hypothesis** that the **median** change in a score in a single group of subjects (but between two **time points**) is zero. ⇨ **paired *t* test**

Wilcoxon rank sum test   ≈ **Wilcoxon matched pairs signed rank test**

window   usually used to refer to a **time period** within which certain events (such as **visit dates**) are allowed to occur. ⇨ **therapeutic range**

window width   the width of a **window** (often measured in days or other units of time)

Winsorise   a method of reducing the influence of **outliers**, typically used when calculating the **sample mean**. Each of the smallest n values is set equal to the next largest value (the $n + 1$th value); similarly each of the largest n values is set equal to the $n - 1$th value

Winsorised mean   the **sample mean** calculated after **Winsorisation** of the data

withdraw   to stop taking part in a study. Generally the implication is that the subject has stopped before completing the study. This could be for a variety of reasons including death (if mortality is not the **primary outcome** of the study) through to voluntary reasons or **loss to follow-up**

withdrawal   a subject who **withdraws** from a study

within groups   usually used in describing the estimate of variation (the **variance**) of data that is due to differences in subjects, not differences in groups (for which, ⇔ **between groups**). ⇨ **between subjects**

within groups sum of squares   a measure of **variability** (by the method of the **sum of squares**) within treatment groups (or within other **strata**). ⇔ **between groups sum of squares**

within groups variance   ≈ within group; this makes it explicit that it is the variance that is being considered

within groups variation   a less formal term for within groups variance

within person   ≈ within subjects

within study   in meta-analyses this is used to describe the variation that is due to differences between subjects, not between studies (for which, ⇔ between study)

within study variance   ≈ within study. This makes it explicit that we are referring to the variance within studies (between subjects)

within study variation   a less formal term for within study variance

within subjects   usually used to refer to the variation in values of repeated measurements in the same subjects

within subjects comparison   situations where subjects are observed, at least twice, under different conditions and the difference between those two occasions is considered. ⇨ paired comparison, paired data

within subjects effect   effects that are observed as changes in individual subjects rather than as differences between groups. ⇨ change from baseline

within subjects study   a study that does not have a control group (but may still be a comparator study, for example if it were a crossover study)

within subjects sum of squares   a measure of variability (by the method of the sum of squares) of repeated measurements in the same subject. ⇔ between subjects sum of squares

within subjects variance   this is similar to measurement error. It is the variance of repeated measurements taken on the same subject

within subjects variation   a less formal term for within subjects variance

within treatment   ≈ within groups

workup bias   any systematic differences between early results from studies and later results caused by subjects or investigators improving their skills at using equipment, becoming more familiar with study procedures, etc. ⇨ learning curve

worst case   when data are missing, a very pessimistic view may be taken on what the data might have been; this is referred to as the worst case. Pessimism in this situation usually means introducing a bias towards the null hypothesis

worst case analysis   analysing datasets by making the least favourable assumptions. This includes imputing missing values as 'bad' results (whatever the context), using crude methods of analysis rather than more sophisticated methods, etc. ⇔ best case analysis

written assent assent that is confirmed in writing. ⇔ **oral assent**.
⇨ **written consent**

written consent consent that is confirmed in writing. This is more
common in clinical trials than **oral consent**, **oral assent** or **written assent**

x   the symbol usually used to denote a variable in a **univariate** situation. ⇔ *y*

x **axis**   the horizontal axis on a graph, **scatter plot**, etc. (Figure 34). ⇔ *y* **axis**

x **coordinate**   the value of *x* in a pair of *xy* **coordinates**

x **variable**   ≈ **independent variable, covariate.** ⇔ *y* **variable**

xy **chart**   ≈ **scatter plot**

xy **coordinate**   the data values (denoted *x* and *y*) that form a pair to be plotted on a **scatter plot** or other similar graph *y*   the symbol usually used to denote the **response variable** in a **regression model** or in **analysis of variance**. ⇔ *x*

y **axis**   the vertical axis on a graph, **scatter plot**, etc. (Figure 34). ⇔ *x* **axis**

y **coordinate**   the value of *y* in a pair of *xy* **coordinates**

y **variable**   ≈ **response variable, dependent variable**. ⇔ *x* **variable**

Yates' continuity correction   ≈ **Yates' correction**

Yates' correction   an adjustment made in the calculation of the **chi-squared statistic** in **two-by-two tables** that is often used in **small samples**. It is used in some other calculations too, but almost always in the case of small sample sizes

Youden square   an experimental design similar to a **Latin square** but where not all treatments occur in all rows or columns of the design

z **axis** when three variables are being plotted on a graph (like an **isometric graph**, Figure 16), the concept of a z axis is usually introduced to refer to the third variable (after x and y). If more than three variables are being considered, then they are not usually given names. ⇨ x **axis,** y **axis**

z **distribution** ≈ **Normal distribution**

z **score** a value from a **standard Normal distribution** (that which has **mean** zero and **variance** equal to one)

z **statistic** the calculated value of z in a z **test**

z **test** a statistical **significance test** concerning the mean of a population when the **variance** is known. The situation that the variance is known is not usual; when it is not known the t **test** is used instead

Zelen's randomised consent design a design that combines **randomisation** with **consent**. Subjects are randomised to one of two treatment groups. Those who are randomised to the **standard treatment** are all treated with it (no consent to take part in a study is sought). Those who are randomised to the **experimental treatment** are asked for their consent: if they agree they are treated with the experimental treatment; if they disagree they are treated with the standard treatment. An alternative is that those randomised to the standard treatment may also be asked if they accept that treatment; again, they are given their treatment of choice. This latter case is described in Figure 10 (≈ **flow diagram**). In either case, the analysis must be based on the treatment to which patients were randomised, not the treatment they actually received. ⇨ **Wennberg's design**

Zelen's single consent design ≈ Zelen's randomised consent design